SIR HUBERT WILKINS

Fact Finder and Truth Seeker

Saskia Praamsma and Matthew Block

SQUARE CIRCLES

SIR HUBERT WILKINS
Fact Finder and Truth Seeker
by Saskia Praamsma and Matthew Block

ISBN: 978-1-7344621-5-9

Cover, interior, and maps: Syrp & Co.

This book reproduces (with the addition of a few images) the summary of Sir Hubert Wilkins' life that appeared in *The Urantia Notebook of Sir Hubert Wilkins,* published by Square Circles Publishing in 2015.

Abbreviations for photo credits:

BPCRC_W	Ohio State University, Byrd Polar and Climate Research Center Archival Program, George Hubert Wilkins Papers
MR:	Collection of Mike Ross, now housed at BPCRC
UCA	University of Central Arkansas, Archives and Special Collections
AWM	Australian War Memorial, Canberra
SPRI	Scott Polar Research Institute, University of Cambridge
UBF	Urantia Book Fellowship website

SQUARE CIRCLES PUBLISHING
www.SquareCirclesPublishing.com

Wilkins was a two-sided individual: the man the world knew and the man who only his most intimate friends knew. He had a deeply spiritual side in the broadest possible cosmic sense. He had inner experiences that he confided to scarcely anyone.

—*Harold Sherman, co-author of* Thoughts Through Space

CONTENTS

MAPS

THE NORTH POLE

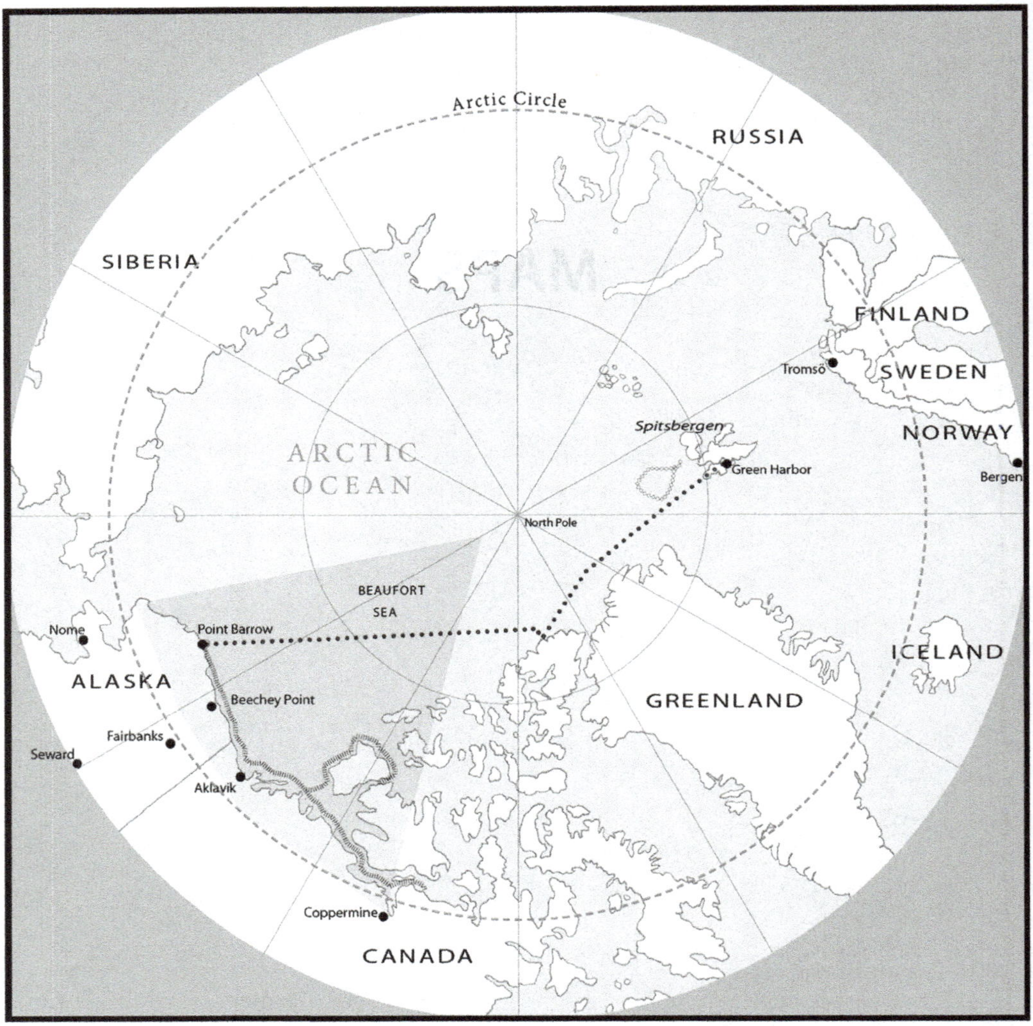

Canadian Arctic Expedition, 1913-1916
Path of historic flight, April 16, 1928
Route of the *Nautilus,* 1931
Area covered in the search for Levanevsky

AUSTRALIA

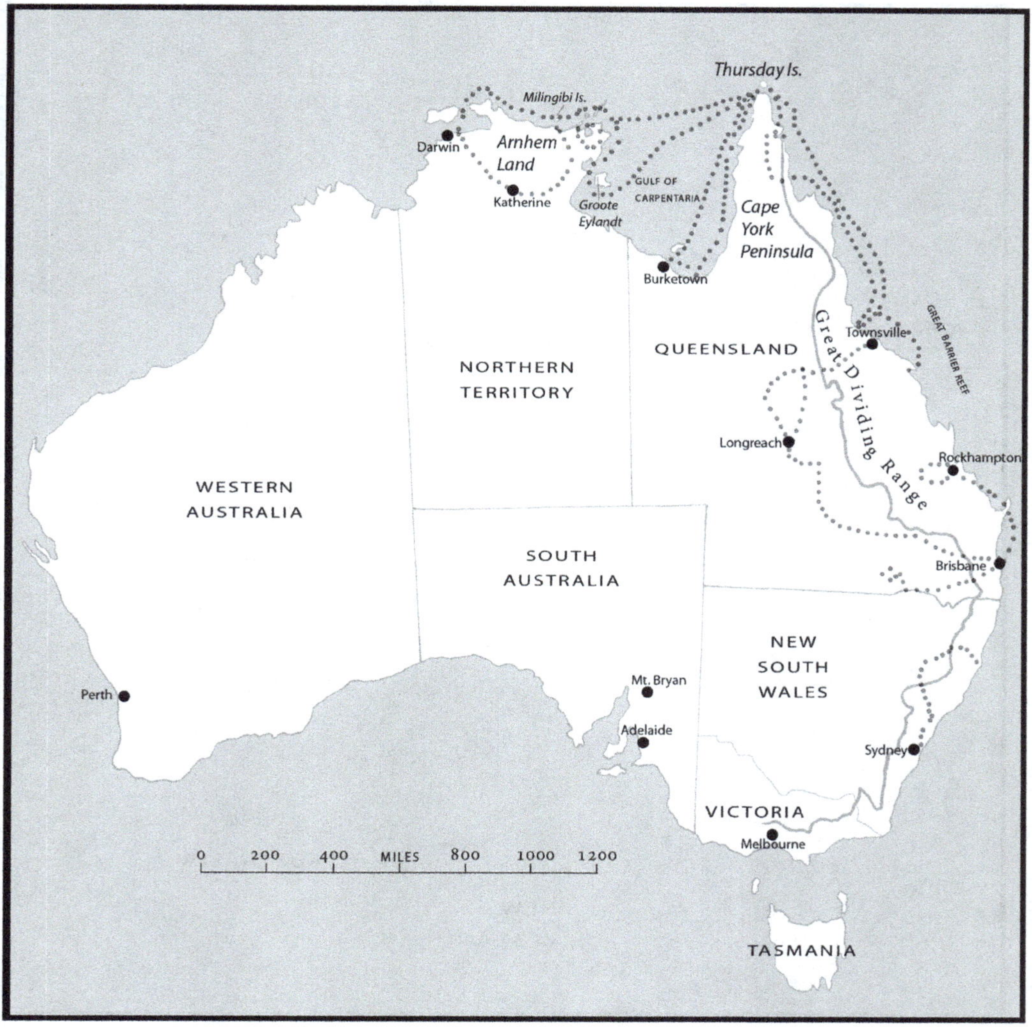

Areas covered by Wilkins during the Australia and Islands Expedition 1923-1925

THE SOUTH POLE

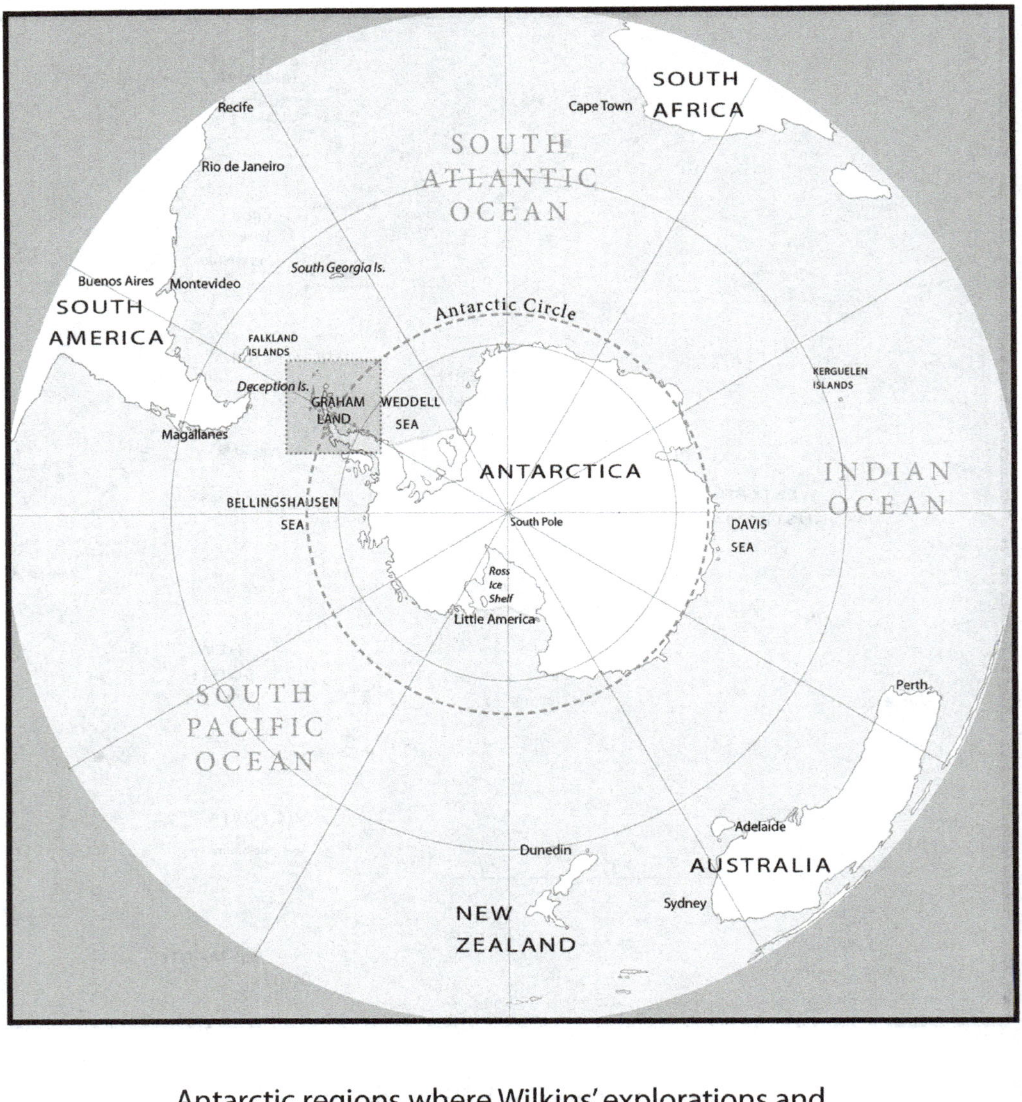

Antarctic regions where Wilkins' explorations and activities took place, 1921-1939

Area enlarged on opposite page

PIONEERING ANTARCTIC FLIGHTS

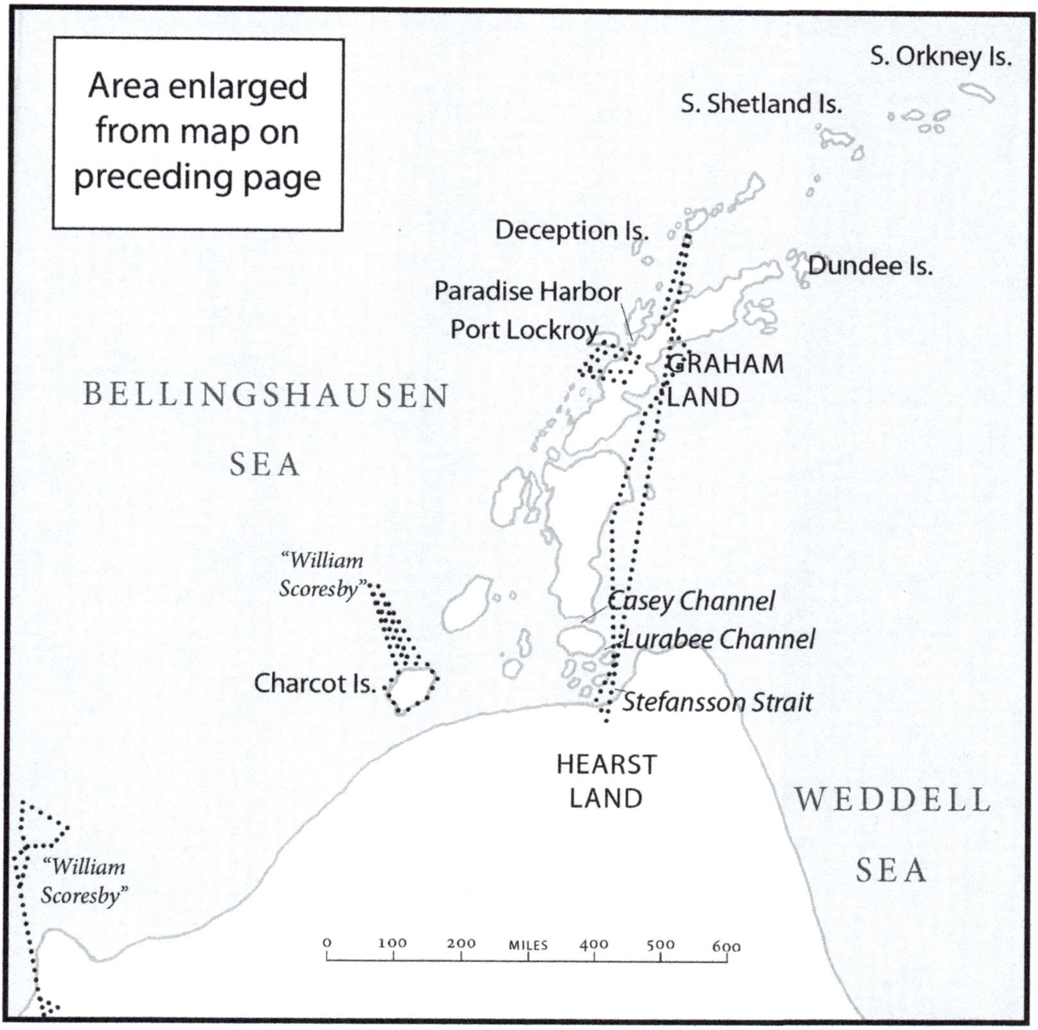

•••••• Wilkins' main Antarctic flights and claimed discoveries, 1928-1939

WILKINS
THE FACT FINDER

THE FACT FINDER

The Wilkins homestead [BPCRC_W]

EARLY YEARS IN MT. BRYAN

George Hubert Wilkins was born on October 31, 1888, into a pioneer family of livestock farmers in Mt. Bryan East, a settlement 120 miles north of Adelaide, in South Australia. He was the youngest of thirteen children of Harry and Louisa Wilkins. Both sides of his family had emigrated from England: Harry's parents, William and Mary Ann Wilkins, arrived in 1837 on the *Emma*; Louisa and her parents, George and Elizabeth Smith, sailed to Australia on the *North* in 1858 when she was sixteen. Living in the outback, George attended a one-teacher bush school and learned, from his father

Harry and Louisa Wilkins [BPCRC_W]

and older brothers, to shear sheep and pursue kangaroos on horseback. As a teenager he became acquainted with aborigines, hunted and camped with them, and learned of their vast understanding of the physical and spiritual realms. Wilkins wrote: "Early in life I was steeped in what we like to call adventure. Before that, it was in the blood and bone of our family—not only adventure as related to the physical body but intellectual as well."[1]

A severe drought beginning in 1901 and lasting four years made a permanent impression on young Wilkins; he was forced to watch his family's animals slowly die of thirst and starvation and many of his once-affluent neighbors desert their parched, dust-ravaged farms. When relief finally came in the form of heavy rains, the farmlands became lush again, providing abundant feed for livestock. But sheep and cattle available for restocking sold for exorbitant prices. And what if another dry spell came upon them? Wilkins reasoned that if the movements of the world's atmosphere could be predicted, farmers could prepare for droughts by stockpiling feed and temporarily relocating their animals.

This experience directed the future course of Wilkins' life. He determined to devote his life to studying the weather and predicting its patterns.

ADELAIDE AND SYDNEY

[MR]

In 1905 the Wilkinses sold their farm and moved to Adelaide, where George studied mechanical engineering at the University of Adelaide and electrical engineering at the Adelaide School of Mines. He also attended the Elder Conservatorium of Music where he was trained in singing as well as the organ and some orchestral instruments. He preferred reading—Darwin, Huxley, Haeckel, Paine and Emerson—to spending time with friends, reasoning that "when there is so much condensed knowledge in books, it seemed a waste to use up time with others of my age who could give me little more than I had myself."[2]

While fixing a generator for a motion picture troupe, George became interested in the film projector and asked the operator to show him how to use it. This was the start of Wilkins' love of filmmaking. From then on he spent all his spare time with the troupe, learning how to make movies himself. He then toured for eighteen months through South Australia and Victoria, as lighting operator for a carnival of singers, acrobats and dancers.

When the tour ended, in late April 1909 George stowed away on a ship to Sydney, where he first worked as a tent cinema operator and later as a theater manager and cinematographer. He produced short action films, using friends and colleagues as actors.

Wilkins as a young filmmaker [BPCRC_W]

But weather study and forecasting remained his primary interest. At twenty he laid out a plan for his life: "I would travel, study and work in the out-of-the-way places of the world for twenty years, then spend the next twenty making use of the information gained to build up a comprehensive international weather service that would adequately include the polar regions."[3] He did not seek meteorological work in Adelaide as he wanted to approach the subject from a global not a local point of view.

LONDON AND ABROAD

In February 1912 the Gaumont Company, the foremost motion picture outfit in Europe, invited Wilkins to come and work for them in London on the condition that he arrange his own passage. Without stopping to see his family in Adelaide, Wilkins sailed to Europe, traveling via Western Australia, the East Indies, North Africa and Italy. He paid his way with his savings and by selling his photographs and travel stories.[4]

In England George worked as a photojournalist for Gaumont and as a reporter on the London *Daily Chronicle,* taking pictures on horseback. His jobs took him around the British Isles and Europe. In the process he became interested in flying and was thrilled when pilot Claude Graham-Waite took him up for the first time, from London's Hendon Aerodrome, in 1910. Wilkins immediately began flying lessons and mingling with aviators. One was Lieutenant Porte of the Royal Navy, who invited him to strap himself and his equipment onto the fuselage of his new plane, to shoot film from the air. While the ride was terrifying and the resultant film unusable, it marked one of the first attempts at aerial motion pictures.

THE BALKAN WAR

Wilkins' first major assignment came in 1912 when Gaumont sent him to Constantinople (Istanbul) in Turkey to film the First Balkan War* from the Turkish side. He joined a 28-member press corps headquartered at the renowned Pera Palace Hotel, sharing servants, horses and provisions with two other journalists, one of whom was Bernard Grant, of the London *Daily Mail.* The Turks were losing the war; they did not welcome news reports about their setbacks and therefore thwarted the reporters' efforts to send out stories. Many of Wilkins' sensational accounts and pictures were dispatched by bribing local drivers and tugboat captains to get the information out of Turkey. During this period Wilkins dressed in Arab clothes, under which he hid his film.

After filming the Battle of Lule Burgas**—possibly the first time in history that a battle was so recorded—Wilkins was arrested by Bulgarian troops but escaped, dodging bullets until he reached a village where he was able to sneak out his report to Constantinople from where it was sent on to London. Soon after, he and Grant were arrested by the Turks and held for ten days for illegally taking pictures.

*Bulgaria, Greece, Serbia and Montenegro formed the Balkan League to fight to free their ethnic populations from Ottoman Turkish rule.

**The principal battle of the First Balkan War, in which the Kingdom of Bulgaria forced the Ottomans to retreat to a point just 30km from Constantinople.

Ottoman troops in retreat from Lule Burgas [BPCRC_W]

Prince Aziz Pasha, a member of the ruling Ottoman family with whom Wilkins had become acquainted in Constantinople, wrote letters of courtesy which led to the reporters' release. In apology, the Minister of Defense provided Wilkins and Grant a two-week stay in a luxurious harem, during which time the war came to an end.[5]

THE WAYWARD BALLOON

Returning to London just before Christmas 1912, after three months in Turkey, Wilkins was commissioned by Gaumont to film a seasonal promotion for Sandow's Cocoa Company, in which Santa Claus would parachute from a balloon over Hyde Park and hand out chocolates to children. The Spencer brothers, owners of a balloon used for demonstrations all over England, were hired together with a daredevil named Captain Penfold who would play Santa. At the last minute, instead of filming the event from the ground as planned, Wilkins decided to join Frank Spencer and Penfold in the air to add weight and get good aerial shots.

Once airborne, the craft immediately headed in the wrong direction and became engulfed in thick clouds. At the first sight of ground Captain Penfold jumped out, causing the gas-release cords to get tangled and unbalancing the balloon, which went sailing towards the North Sea. During the night the balloon rose to 22,000 feet, its occupants hungry, scared, and freezing. By morning the winds had changed and the two were relieved to find themselves over the English Channel headed toward the coast. The balloon, however, was dangerously low over the water, skimming it from time to time, and one severe jolt caused Gaumont's uninsured movie equipment to tumble overboard. Eventually they crash-landed in a farmer's field, but both men survived. Captain Penfold, meanwhile, had landed on his head in a field thirty-five miles north of Hyde Park, but survived as well.

Gaumont was furious with Wilkins over the loss of expensive equipment, with not a foot of film to show for their efforts.

Captain Penfold and Wilkins (black hat) prepare for an airborne mission [MR]

COCOA PRODUCTION IN TRINIDAD

Plantation workers in Trinidad photographed by George Wilkins [BPCRC_W]

A complete change of scenery was next in store for George Wilkins in February 1913, when he was sent by Gaumont and the Cadbury chocolate company to the Caribbean island of Trinidad, to shoot a short publicity film about the cocoa industry. Staying in a pleasant home with servants, Wilkins spent his days at the picturesque tropical cocoa plantations recording routine operations and returning home in the evening to develop the film. His images captured the tall cocoa trees grown in the shade of an even taller Mother Tree, and the colorful laborers picking, fermenting and drying the beans and then roasting them like coffee.

The native workers were used as indentured slaves and kept away from the white men according to strict rules. When Wilkins wanted to film the women doing their traditional dancing, which only husbands were permitted to see, he recruited the help of the overseer's sister who arranged for him to dress as a woman and enter with his camera while the men were working elsewhere. As Wilkins filmed, the singing and dancing grew so loud and frenzied that the men heard it and one husband came over to investigate. Seeing his wife with another man—a transgression of the moral code—the native slashed off her ear, the penalty for such a violation. When the others later determined that the woman was innocent because only the camera had seen her, her honor was restored and she proudly displayed her wound.[6]

Wilkins traveled to the other islands of the Lesser Antilles, filming much of their scenery and many industries. The film he shot for Cadbury's was hand-colored by Gaumont in London and released in 1914 as *Food of the Gods.*

THE ARCTIC WITH STEFANSSON

Upon his return to London in April 1913, Wilkins found a cable from Canadian polar explorer Vilhjalmur Stefansson inviting him to join his upcoming Canadian Arctic Expedition as photographer. The goal of the expedition, projected to take five years, was to survey a 40,000-square-mile area from the Arctic coast of Canada to the North Pole, where a massive continent was believed to exist, and to claim that land for Canada. This appealed to Wilkins because it

Surviving on the ice with Stefansson and expedition members [BPCRC_W]

Wilkins showing films to Eskimos [BPCRC_W]

gave him the opportunity not only to film the region but to observe Arctic weather patterns. Stefansson was a veteran of two previous Arctic expeditions and had learned from the Eskimos how to survive on the ice.

In June Wilkins joined the mother ship, the *Karluk*, in Victoria, British Columbia, with a team of scientists on board and Captain Bob Bartlett at the helm. They sailed to Nome, Alaska, where Stefansson joined them, and from there to Point Barrow, the northernmost outpost on land. Four other ships were also involved in the mission.

Disaster struck on September 20 when the *Karluk*, heading north into the Arctic Ocean, became trapped in the ice fifteen miles offshore. Stefansson, Wilkins and two others as well as a team of dogs disembarked to seek help. They never found the *Karluk* again. The ship, they later learned, had drifted with the ice for fifteen months before sinking; twenty-five survivors broke up into small groups going in different directions, and eleven died. Meanwhile, Stefansson and his party lived on seals, bears, ducks, owls, loons, ravens, foxes and squirrels. Stefansson taught the men how to drive dogs, carry packs, and sleep naked in bags made of reindeer skin inside a self-made snow house. Eventually they were able to reunite with expedition members from the other four ships, replenish their equipment and continue their work.

On this trip, which lasted three years for him, Wilkins took hundreds of photos. He enjoyed the life of an explorer and picked up many skills from the scientists, including collecting and experimenting with biological specimens, taxidermy, and map-making. In all, four new islands were discovered and many regional maps revised. Stefansson had made Wilkins second-in-command of the expedition, and his colleagues named Wilkins Bay, Wilkins Strait, and Wilkins Point after him to show their respect. Stefansson described Wilkins as "easily my best man"[7] and began to regard him as a protégé. The Eskimos taught Wilkins further survival techniques, and he in turn transcribed their music and songs for ethnologists using his own brand of musical notations.

While traversing much of the terrain on foot, Wilkins observed smooth patches of ice that would make perfect landing strips. He realized that it would be much more efficient to conduct future polar exploration with airplanes and submarines.

THE GREAT WAR

When Wilkins returned to civilization in 1916 he learned that his father had died and that the Great War was in progress in Europe. Seeking to serve with the Australians, he sailed from North America to Australia via the wartime shipping lanes. While crossing the Atlantic in one of a convoy of ships, he survived a torpedo attack by German U-boats. He then traveled from England on a steamer around the Cape of Good Hope. During the long voyage he completed a report of his Arctic work with Stefansson for the Canadian Ministry of Marine.

In April 1917, after reuniting briefly with his family in Adelaide, where he was welcomed as a minor celebrity and invited to give talks before large audiences, Wilkins was commissioned by the Royal Australian Flying Corps and given the rank of lieutenant. However, upon arrival in England he was declared medically unfit due to color-

Wartime activities photographed by Wilkins at Ypres in Belgium [BPCRC_W]

blindness and frostbite damage to his feet. Instead of flying he was assigned to join the Australian War Records Section, headed by veteran war correspondent Captain C. E. W. Bean, as photographer covering Australians in battle. Wilkins was to accompany Frank Hurley, who had been Sir Ernest Shackleton's photographer on his 1914-1917 Antarctic expedition. The two were given separate assignments: Hurley was to take propaganda shots to drum up support for the war while Wilkins was to produce an accurate visual record for future historians.[8]

In July 1917, Lieut. G. H. Wilkins arrived in Boulogne and was taken to film the Third Battle of Ypres, in Belgium.* He and Hurley, together with a staff, were housed six miles from the front across the border in France, in the village of Steenvoorde. Every day they arrived at the battle scene at sunrise to capture images of front-line activity from the ground as well as from airplanes and balloons; at sunset they returned to their headquarters to develop the film, which

*The Third Battle of Ypres lasted from July to November 1917. It was a failed attempt by the Allies to destroy the German submarine bases on the Belgian coast.

was then dispatched to censors. On September 27, 1918, Wilkins was with the Americans, supported by Australians, when they attacked the Hindenburg Line, the last defense of the German Army, an action that led to the end of the war.

In his eighteen months on the job Wilkins was wounded many times. His bravery under fire, including assisting wounded men, twice earned him the Military Cross. During this time he was promoted to Captain. Australian Commanding General John Monash described Wilkins as "the bravest man I have ever seen."[9] Many of Wilkins' photographs were used in *The Official History of Australia in the War of 1914-1918.*

AFTERMATH OF GALLIPOLI

Wilkins (center) and Bean (to his left) and other members of the Australian Historical Mission having lunch at Hill 60 on the Gallipoli Peninsula [AWM]

Like Stefansson before him, Captain Bean had begun to regard Wilkins as a protégé. The war over, he invited Wilkins and a team of artists and historians to the Dardanelles—the strait separating European from Asian Turkey—to take part in the Australian Historical Mission of 1919. The purpose of the mission was to reconstruct the events of the Gallipoli Campaign* from both the allied and enemy perspectives and to retrieve personal artifacts left behind. Bean described Wilkins as a born leader, noting that "if he used a certain path, climbed a cliff in a particular way, jumped a trench or even went round the left or right of a bush, the rest of us usually did the same."[10]

Wilkins and Bean and their team then journeyed 1,500 miles by rail from Turkey, through Syria, Lebanon, and the Sinai Desert to Cairo, to check up on Australian graves in the Middle East. Wilkins filmed all along the way. In Cairo riots had broken out against Europeans, especially the Imperialist Brits, which Wilkins photographed from a hotel balcony. He and an associate were arrested, but when the Egyptians learned that Wilkins was Australian they set him free, being sympathetic to a fellow victim of "colonial oppression."[11]

ENGLAND TO AUSTRALIA AIR RACE

The First World War had produced many pilots eager to demonstrate their skills and venture into unknown territory. In 1919 the London *Daily Mail* offered £10,000 to the first air crew to successfully cross the Atlantic. Hoping to give Australian aviation international exposure, Prime Minister Billy Hughes proposed the same prize for an air race from England to Australia. He stipulated that the crew had to be Australian and consist of a pilot, navigator and mechanic; that all parts and equipment were to be of British Empire manufacture; that the crew could not take off before September 8, 1919; that the journey had to be completed in thirty days and finished before the end of the year. In the interest of safety, novices were ruled out.

Six crews entered into the competition, which was backed by aircraft manufacturers and fuel companies. One entrant was Valdemar Rendle, who had flown for Britain in the war. He secured a twin-engine Blackburn Kangaroo and took on George Wilkins, Reg Williams

*The Gallipoli Campaign of 1915-16 was an unsuccessful attempt by the British and French to control the sea route from Europe to Russia during World War I. The Allies, which included the Australian and New Zealand Army Corps (ANZAC), were unable to break the Turkish defenses.

The Blackburn Kangaroo at rest, just short of a stone wall in Crete [BPCRC]

and Gar Potts, all experienced pilots and mechanics. Wilkins was designated commander and navigator. Seeing the race as an opportunity for meteorological research, Wilkins brought along a drift indicator, a thermometer, and a humidity meter. He planned to measure wind conditions, cloud formations and tropical turbulence. He also designed his own navigation instruments which would allow them to fly through fog, clouds and darkness—the first known attempt at instrument flying. The *Sydney Sun* quoted Wilkins as saying: "I would not do this as a stunt. I have definite scientific reasons for going." His motive, as he expressed in a letter to his mother on the eve of departure, was "to do such things as will be useful to my fellow man."[12]

On November 21, after many delays that gave other teams a head start, Wilkins' group took off from Hounslow Aerodrome, waved off by Prince Albert and Winston Churchill. Bad weather soon forced them to land at a military field near Paris where they were grounded for three days. Flying through a blinding snowstorm over the Alps, the controls froze; the pilot maneuvered the plane down by switching the engine on and off, narrowly missing the ruins of an ancient Roman aqueduct as they careened into an empty hangar at the end of an airstrip in Frejus. With the help of expat British mechanics they got the Kangaroo back in the air the next morning, but within minutes the engines cut out and they made an emergency landing. Upon examination they found that an ignition wire had been tampered with. Sabotage?

After a thorough inspection they set off again and made it to Pisa, arriving drenched by the heavy rain that poured into their open cockpit. Attempting to take off again, the Kangaroo became bogged down in mud and had to be dug out, causing another day's delay. (News reports meanwhile told them that another team, the Smith brothers, had reached Burma.) Arriving in Rome on All Saints' Day when everything was closed, the crew had to pass more time until they could load fuel and other supplies. (The Smiths, they heard, had now reached Bangkok.) Next they landed in Crete on more muddy runways. Turkish prisoners were brought in to dig drainage ditches and construct a makeshift airstrip, and once again the Kangaroo set off, this time for Egypt.

About eight miles out, Potts noticed oil gushing from one of the engines, forcing them to turn back. Miraculously, a strong wind prevented the Kangaroo from losing altitude as it sputtered along on one engine. When the coastline was spotted Wilkins looked for a safe place to land but the wind began carrying them in the opposite direction. In desperation Rendle tried to restart the bad engine, but without oil it exploded and the plane crash-landed in a six-foot ditch just short of a high stone wall. Because damage was minimal they planned to continue and ordered a new Rolls Royce engine from England, but before it could be shipped word reached them that the Smith brothers had landed in Darwin and claimed the £10,000 prize.

THE BRITISH IMPERIAL ANTARCTIC EXPEDITION

Still harboring his dream of establishing a series of meteorological stations in the polar regions to aid in weather forecasting, in 1920 Wilkins joined the British Imperial Antarctic Expedition, headed by John Lachlan Cope, who had served on Sir Ernest Shackleton's 1914-1917 Imperial Trans-Antarctic Team as biologist and surgeon.

At that time the Antarctic was mostly unexplored; no plane had flown over it and only Norwegian explorer Roald Amundsen had reached the South Pole (in 1911) and survived. Cope's plans were for a five-year mission to circumnavigate Antarctica by sea, establishing bases inland from where to survey the region by air using twelve aircraft. A team of a hundred scientists and explorers would be employed. Discovered land would be claimed for Britain. The idea of permanent bases fitted in with Wilkins' goals, and he signed on as second-in-command and chief of the aeronautical division.

While preparations were underway, Wilkins took a quick trip back to Australia to help Captain Bean organize their vast collection of World War I photos. While there, he lectured on his plans for permanent weather stations in the Antarctic and predicted that one day wireless communication between the southern regions of the continents—South America, Africa and Australia—would enable forecasting weather months or even years ahead.

Upon return to England he learned that the Royal Geographic Society had refused to sanction the Cope expedition, leaving its leader with few financial backers. Instead of a hundred-member team there would now be just Cope and Wilkins plus two young, inexperienced volunteers—Thomas Bagshawe as geologist and Maxime Lester as surveyor. No aircraft would be used. Wilkins, whose main reason for joining was to survey the area from the air, tried to bow out but Cope convinced him to stay.

The grand plan was reduced to a mission of gathering as much topographical knowledge of the western coastline of the Weddell Sea as possible. They would hitch rides on whaling ships from Montevideo to the whaling center of Deception Island and from there to Paradise Harbor in Graham Land. There they would stay until the end of whaling season. For shore excursions they planned to use a 16-ft open whaleboat and a team of dogs.

When Wilkins met up with Cope in Montevideo, he learned that Cope was deep in debt. The whalers had taken Wilkins' movie equipment as part collateral, and agreed to ferry the group south only on

Lester, Bagshawe and Cope standing in front of their base hut [SPRI]

condition that Wilkins take over as expedition leader. In exchange for their cooperation, Wilkins negotiated a deal to film their whaling operations for publicity.

On January 12, 1921, the crew of four—Wilkins, Cope, Bagshawe and Lester—landed at Paradise Harbor. The landscape of steep, icy cliffs made sledding impossible. Twice Wilkins barely escaped with his life: first when he and Cope, roped together, tumbled down a sharp cliff; and again when their small open whaleboat, having sprung a leak from crashing into underwater rocks during a heavy storm, almost capsized. Wilkins wrote later: "This strange force saved us from certain death. I felt this force many other times in my life. It was inexplicable, quite baffling. It was as if a ghost would walk at my side."[13]

After four months of researching, Wilkins and Cope left Bagshawe and Lester behind, at their own request. Cope earned his passage home by peeling potatoes on a Scottish steamer, never to return to Antarctica. Wilkins sailed to New York determined to come back with an airplane. Regarding his experience with the British Imperial Antarctic Expedition, he said, "I was really not with much of an expedition at all, just four fellows camping in the Antarctic for a few months of summer."[14]

THE SHACKLETON-ROWETT ANTARCTIC EXPEDITION

Back in New York in May, the German Junker Company offered Wilkins two planes for an Arctic expedition, but while he was negotiating he received a cable from Sir Ernest Shackleton strongly advising him not to deal with the Germans so soon after their war with Britain. Instead, he offered Wilkins a British plane if he would agree to join his upcoming Shackleton-Rowett Antarctic Expedition.

Though Shackleton's 1914-1917 expedition to Antarctica had also failed, Shackleton himself came through the experience highly regarded by his team, who called him "the Boss." The new expedition was financed by Shackleton's wealthy school friend, John Quiller Rowett, and would start in September 1921. Using a seaplane and a Norwegian sealer that he renamed the *Quest*, Shackleton proposed to sail completely around the Antarctic continent with frequent inland flights. Wilkins signed on as aviator and naturalist. The London *Daily Mail* promoted the expedition by organizing an essay contest among the Boy Scouts, the winner to serve as cabin boy on the team.*

Wilkins arrived in London in June, and while waiting for the expedition to be outfitted, he practiced flying the new British airships—also called dirigibles—in order to sit for the air navigation exam. In a catastrophic test flight of the R-38—a voyage that would have included Wilkins but for the too-late arrival of a telegram informing him of the departure time—both of Wilkins' trainers and forty-four others were killed. Once again the guiding hand of fate had saved him.

On September 17, the *Quest* set sail from London but the voyage was hampered by bad weather and endless mechanical woes. Several

*Of the ten finalists, Shackleton chose two boys: seventeen-year-old Norman Mooney and eighteen-year-old Aberdeen geology undergraduate James Marr. Mooney left the expedition early due to seasickness; Marr wrote about his experience in his 1923 book, *Into the Frozen South*, and became a polar geologist.

The Quest *trapped in the ice* [BPCRC_W]

of the crew were set ashore along the way due to seasickness, including the photographer whose duties Wilkins took over. To make up time they bypassed their planned stop in Cape Town to pick up the plane and sailed directly to Rio de Janeiro. Once again Wilkins' dream of flying over the Antarctic was shattered.

Rather than kill time socializing with the British colony in Rio for the six weeks that it would take to repair the *Quest*, Wilkins and geologist George Douglas went ahead on a whaling vessel to the island of South Georgia, then the world's largest whaling center, where Shackleton and the group would later join them. Wilkins savored the mountainous beauty of the island and spent six highly enjoyable weeks collecting samples of flora and fauna, returning at night to his tent to cook a meal and prepare his specimens by candlelight. He photographed bird life, snow-capped mountains and tiny fishing settlements.

On January 4, 1922, having received word that the *Quest* was in port, Wilkins arrived on the scene to be greeted with the devastating news that Shackleton had died on board the night before of a massive heart attack. The crew was heartbroken. Buried at the foot of the South Georgian mountains, Shackleton was mourned around the world. Wilkins described him as always cheerful and entertaining, no matter how desperate the situation, and "the most beloved leader of men I ever knew."[15]

Second-in-command Frank Wild attempted against all odds to continue with the original plans, hoping to discover many miles of

Wilkins' tent on South Georgia Island [BPCRC_W]

Members of the expedition at their memorial service for Shackleton [BPCRC_W]

new coast and unknown species. While exploring around the frozen Weddell Sea, the inadequate *Quest* became trapped in the ice. Soon the team faced an even greater dilemma: a massive iceberg was headed their way, spelling certain doom. As the men prepared to abandon ship, the berg inexplicably turned and began drifting in another direction, leaving behind an open passage free of ice for the vessel to escape through. Wilkins again credited the mysterious unseen hand with rescuing them.

With the *Quest* plainly unsuited for polar work and the weather against them as well, the expedition was called off and the crew sailed back to South Georgia, arriving in April. On the voyage back to England, they stopped at remote island groups where Wilkins continued his field work. In the cramped quarters he shared with George Douglas, Wilkins had set up a makeshift darkroom, a laboratory, and storage units for specimens, and made copious notes. Ship's navigator Douglas Jeffrey described Wilkins as "a painstaking scientist and an excellent companion . . . always cheery and always working."[16]

UNDERCOVER IN RUSSIA

Back in London in 1922, George Wilkins' reputation as explorer and war photographer made him a popular guest at social functions with government and military groups. On one of these occasions he was invited to travel to Russia on a fact-finding mission and to arrange meetings with Russian political leaders.

Seven years of constant warfare in Russia—World War I, the Revolutions of 1917, and the Civil War of 1918-1921 between the Red and the White Armies—had decimated the peasants' grain stores. The situation was aggravated when drought conditions in 1921 produced a complete crop failure, resulting in widespread famine. Millions were dying; there were even reports of cannibalism. Unable to deal with the runaway disaster, the Bolsheviks reluctantly agreed to accept help from foreign charities.

The famine gave Wilkins a pretext to go into Russia. His cover was as a photographer and reporter for the Society of Friends (the Quakers), who were doing volunteer work with their Famine Relief Fund and hoping to raise money by providing reports and images of the conditions. Wilkins was accompanied by Lucita Squier, a Quaker mission worker and one-time Hollywood scriptwriter. Her fiancé, *New York Post* reporter and socialist sympathizer Albert Rhys Williams, was stationed in Moscow and had befriended Vladimir Lenin and other Bolshevik leaders. Squier was unaware that Wilkins was on a mission of espionage.

Starting out on November 7, 1922, the pair traveled by train through France, Switzerland, Austria and Czechoslovakia, and by car through Poland to the Russian border, then by train to Moscow. Here they were joined by Laurette Citroën, daughter of the French auto magnate Andre Citroën, who was devoting her life to the poor. The trio spent three months crossing a bitterly cold Russia, reporting, filming, and distributing food which was brought overland from England and transferred onto horse- and bullock-drawn carts. Taking precautions against disease, they traveled by train and droshki, on horse- and camelback, and slept in barns and open shelters.

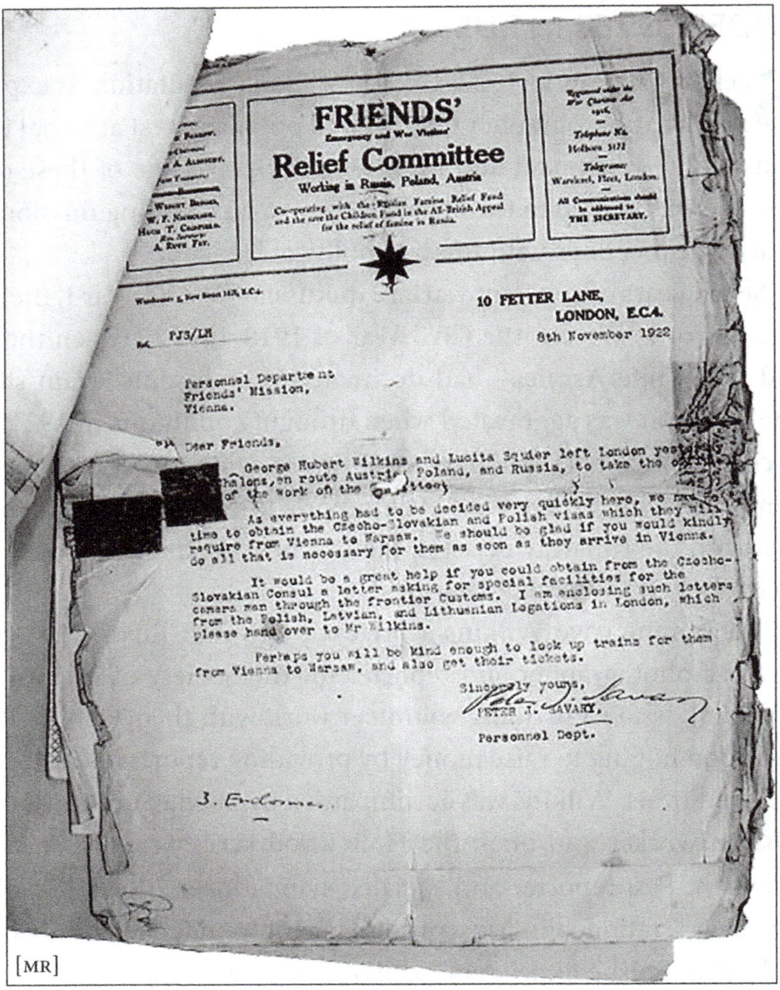

FRIENDS'
Emergency and War Victims'
Relief Committee
Working in Russia, Poland, Austria

Co-operating with the Russian Famine Relief Fund and the Save the Children Fund in the All-British Appeal for the relief of famine in Russia.

All Communications should be addressed to THE SECRETARY.

10 FETTER LANE,
LONDON, E.C.4.
8th November 1922

Ref. PJS/LM

Personnel Department
Friends' Mission,
Vienna.

Dear Friends,

George Hubert Wilkins and Lucita Squier left London yest[erday]
[redacted]halons, en route Austria, Poland, and Russia, to take the
of the work of the

As everything had to be decided very quickly here, we
time to obtain the Czecho-Slovakian and Polish visas which they will
require from Vienna to Warsaw. We should be glad if you would kindly
do all that is necessary for them as soon as they arrive in Vienna.

It would be a great help if you could obtain from the Czecho-
Slovakian Consul a letter asking for special facilities for the
camera men through the frontier Customs. I am enclosing such letters
from the Polish, Latvian, and Lithuanian Legations in London, which
please hand over to Mr Wilkins.

Perhaps you will be kind enough to look up trains for them
from Vienna to Warsaw, and also get their tickets.

Sincerely yours,
PETER J. SAVARY,
Personnel Dept.

3. Enclosures.

[MR]

A 35-minute film, "New Worlds for Old: Quaker Relief in Stricken Europe," with G. H. Wilkins credited as cameraman, was produced for the Society of Friends.[17] Witnessing again the results of a severe drought reinforced Wilkins' desire to establish permanent weather stations around the world.

In December, days before the creation of the Soviet Union, Rhys Williams arranged a meeting between Wilkins and Lenin. No record of the meeting exists, but Wilkins wrote about it much later,[18] revealing that Lenin had admitted overestimating the willingness of the working classes to embrace Communism.

AUSTRALIA AND ISLANDS EXPEDITION

When he handed over his specimens from the Shackleton-Rowett Antarctic expedition to the British Museum of Natural History, Wilkins had offered to do "anything possible" to help the museum build its collections. In early 1923, while still in Russia, he received surprising word from the museum that funding had been granted for a "Wilkins Australia and Islands Expedition"—a mission he had neither requested nor expected. Seeing the value of the work, he agreed to the assignment.

Wilkins and the expedition team in Queensland [BPCRC_W]

For tens of thousands of years the aborigines of Australia had lived in harmony with the forbidding terrain, but the fertile areas were in danger of destruction by the actions of the white man, who had bulldozed the land to provide grazing meadows for cattle and sheep. The balance was further upset by the introduction of non-native rabbits, foxes, cats, camels and cacti, which competed with the indigenous animals and plants. The Australian press applauded the Wilkins mission to collect native specimens while there was still time.

Arriving in Australia, Wilkins had difficulty forming his expedition party. University professors discouraged students from joining as there was "no money to be made in expeditionary work."[19] Wilkins was disappointed in his countrymen for their "absence of express desire for culture and for higher things"[20] and their unwillingness to make sacrifices in the name of science. Eventually he formed a team composed of botanist J. E. Young, bird expert Oscar G. Cornwall, and mammalogist Vladimir Kotoff.

Setting up headquarters in Brisbane in the spring of 1923, the foursome planned to travel up through central Queensland to Cape York, collecting mammals on both sides of the Great Dividing Range, and from there to the islands and Arnhem Land in Northern Territory. The team planned to use various modes of transportation: a Model-T car; a fully-equipped railway carriage that hooked onto regular trains; steamers and flat-bottomed boats; and horses.

Their schedule was to rise early in their camp and visit the traps they had set, have breakfast, hunt, bring in specimens and prepare them, have lunch, hunt for several more hours, have dinner, prepare their afternoon specimens, then set out looking for nocturnal animals until midnight.

In December they returned to Brisbane to sort their specimens and make plans for the second year. Wilkins split the team up to work in different locations and when he set off again he was mostly alone.

While in Arnhem Land, he befriended a small, isolated tribe of cannibalistic aborigines, some of whom spoke pidgin English, and for the next two months he camped with them, going wherever they went. Because of police brutality that included shooting black men on sight, the natives were resentful of white men, but Wilkins stayed busy collecting his specimens, preserving and cataloguing them rather than eating them. This the natives found bizarre, even crazy. According to their moral code they would not hurt someone who was mentally unsound, so he was safe. They fed him, protected him, even offered to build him a proper shelter, but Wilkins assured them

Aborigines photographed by Wilkins in Arnhem Land [BPCRC_W]

he was content living as they did, realizing this was the chance of a lifetime to observe their ways and customs and capture them in photographs. While Wilkins was with them they refrained from cannibalism, living on honey and hearts of palms.

On the island of Milingimbi, where the Methodist Overseas Mission had established a settlement, Wilkins endeared himself to a tribe of aborigines by taking part in games with the mission children, playing the organ in church on Sunday, refusing the offer of their wives, and not preaching to them. He simply joined in their ways and earned their trust; some would come up to him while he was sorting his specimens and put their arms around his neck, silently pressing their faces against his as a show of deep affection.

At the end of two and half years, Wilkins had collected some 5,000 specimens of hundreds of species—mammals, reptiles, birds, fish, insects and plants. He took many photos, gathered boxes of fossils and mineral samples, human skeletons and dinosaur bones. Two new species, a rock wallaby and a small lizard, were named after him. On the voyage back to England in late 1925 he dictated his notes to a stenographer, which were published as the 1928 book, *Undiscovered Australia*. His revelations of random killings of aborigines by the police, as well as the cannibalistic practices of the natives, were coolly received by many of his fellow Australians.

With Wilkins going only £10 over budget, the museum, expecting to pay double his original estimate, gave him a grant equivalent to $15,000, which he used to finance his next expedition. Proceeds from his lectures, articles and book royalties he donated to the British Museum.

THE WILKINS-DETROIT MISSION

Although Admiral Peary had reached the North Pole in 1909, the region was still basically unexplored, with dotted lines representing possible land on maps and atlases. Wilkins, interested as always in establishing Arctic weather stations, intended to fly between Point Barrow and the North Pole, then possibly cross the North Pole over to Spitsbergen, a Norwegian island bordering on the Arctic Sea. Nobody had attempted this before and many said it could not be done.

Through Vilhjalmur Stefansson's influence the American Geographical Society endorsed Wilkins' plans, paving the way for financial backers to come forward. Stefansson also hooked Wilkins

The Alaskan *ready to take off on its mission* [BPCRC_W]

up with the Detroit Aviation Society, a group of powerful businessmen who wanted to promote Detroit as not only the center of auto production but of aviation as well. The Society agreed to support Wilkins for one year, as did the North American Newspaper Alliance (NANA) and *The Detroit News*, which contributed a sizable sum in exchange for exclusive news and pictures.

With these funds, Wilkins bought two planes designed by aviation pioneer Anthony Fokker. Stefansson suggested Ben Eielson, a fearless stunt and airmail flyer, as chief pilot. Any land found would be claimed for the United States, and because this caused a dilemma for Australian Wilkins, an American pilot, Tom Lanphier, was brought on as second-in-command. Wilkins' last weeks in Detroit were spent, as he described it, "begging for money"[21] via lunches, dinners, and banquets organized by his backers.

The planes, named *Alaskan* and *Detroiter*, were transported by sea from San Francisco to Seward, Alaska, then by rail to Fairbanks. There, on March 11, 1926, the mission was launched—and tragically, when reporter Palmer Hutchinson was killed instantly by a propeller on the first test flight. Several other test flights in Fairbanks ended in crashes and narrow escapes due to engine trouble.

Meanwhile, gasoline and equipment were being ferried so slowly by dogs to the northern base of Point Barrow, five hundred miles away, that Wilkins and Eielson—waiting for the damaged *Detroiter* to be repaired—decided to freight gasoline up themselves in the *Alaskan*. This enabled them to do their first work of exploration. With Eielson at the controls, Wilkins attended to instrument readings, checking maps, noting geographical and geological observations, and writing notes to Eielson who could hear nothing over the noise of the engines. They were able to add thousands of square miles of Alaskan territory to the map. A possible future landing strip—a large frozen lake ten by thirty miles high in the Endicott Range—was sighted, as well as plentiful caribou for food should they ever be marooned.

Wilkins found, at Point Barrow, basically the same population that he had left behind there in 1913: a handful of Europeans, a couple hundred Eskimos and about eight hundred dogs. Mail was still delivered three times a year by dog teams and the nearest phone was in Fairbanks. Now, via a short-wave radio installed on the plane, Wilkins was able to establish two-way wireless communication from Barrow to the civilized world.

Wilkins and Eielson made fourteen runs between Fairbanks and Barrow before they wrecked the *Alaskan* beyond immediate repair, forcing them to turn to the unstable *Detroiter*.

Around the same time Wilkins arrived in Alaska, the *New York Times* had announced sponsorship of the Amundsen-Ellsworth Expedition. Roald Amundsen, together with American explorer and financier Lincoln Ellsworth and designer and pilot Umberto Nobile, planned to fly an airship, the *Norge*, from Spitsbergen over the North Pole to Alaska. This would be the first attempt in history to fly over the North Pole. Not to be outdone, American Admiral Richard E. Byrd quickly arranged his own expedition to fly from Spitsbergen to the North Pole. On May 9 Byrd took off and claimed to have reached and circled the Pole—a claim that remains controversial to this day.[22]

Early in the morning on May 13 Wilkins and his team awoke in Point Barrow to the glorious spectacle of the airship *Norge* floating majestically over the horizon. It had left Spitsbergen two days earlier and had flown over the Pole, heading for Alaska. The group below waved and cheered, but later those above said they had not seen them. Wilkins was thrilled when the dirigible came into view, since it proved that such a vessel could be used for research work in the Arctic. He was able to scoop the rest of the press with his wireless report of the successful flight.

Wilkins and his team made some further discoveries using the *Detroiter*, but by early June setbacks including severe weather forced them to suspend the mission until the following year. It remained to be discovered if there was land in the Arctic Ocean between Point Barrow and the North Pole, and Wilkins still intended to fly from Alaska to Spitsbergen.

THE SECOND DETROIT MISSION

Back in the States, Wilkins returned to the lecture circuit to raise funds for the continuation of his expedition. In February 1927, he and Eielson, with two Stinson planes and a second pilot, Alger Graham, arrived back in Fairbanks to try once more to fly over the polar cap. A third pilot, Joe Crosson, came on board ferrying a NANA journalist to give up-to-the-minute reports.

Wilkins inside a snow house [BPCRC_W]

On March 29, while exploring the region near Point Barrow, Wilkins and Eielson had engine trouble and were forced to set down on the ice, a historical first, proving that arctic ice could be used as a landing strip. By chipping two small holes in the ice, detonating a stick of dynamite inside one and placing an echo sounder in the other, Wilkins learned that the ocean below was over three miles deep, indicating that there was no land in the vicinity.

After several attempts to get the plane back in the air, they ran out of fuel. Swerving and pitching, Eielson brought the plane down, clipping a small ridge that spun the plane to a complete rest, just before it would certainly have plowed into a massive ice ridge—the unseen hand again reaching out to save them. Here, sixty miles from land on an ice mass drifting eastward, they were stranded for a week in a blizzard, eating their ration of nuts, raisins, biscuits, chocolates, malted milk tablets and pemmican, and drinking snow melted in thermoses inside their sleeping bags. In Barrow, radio operator Howard Mason had picked up faint signals from the Stinson's wavelength reporting a forced landing on ice and a damaged plane.

After the sixth day, they began walking. They wore Eskimo furs and reindeer skin boots and had discarded everything except their instruments, rifles and food, which they carried on their backs in their sleeping bags. Eielson's fingers were frostbitten, so Wilkins alone performed the two-hour nightly ritual of making snow houses and turning their Eskimo clothes inside out to scrape them of snow and frost.

A close call came for Wilkins one day when, loaded down with his heavy bundle, the ice gave way beneath him and he fell into the chilly water. His clothes soaked and frozen, he saved his own life by crawling out, throwing off his pack, stripping naked and dancing around wildly to keep his circulation going. He then frantically rubbed his wet clothes on dry ice to squeeze the water out. Eielson, witnessing the spectacle and horrified that Wilkins had lost his mind, felt they were now both doomed.

After two weeks, stumbling, slipping and staggering on hands and knees, they reached Beechey Point, about 125 miles east of where they had abandoned the plane. Here Eskimos, on the lookout via messages dropped from the air by second pilot Graham, rushed to meet them with dogsleds. Later Graham found them and flew them back to Point Barrow.

Eielson, ever confident that Wilkins would lead them to safety, wrote later: "I believe [Wilkins] knows more about the ice than any man living, and on trail walking I have never seen his equal." The Eskimos in Point Barrow named Wilkins "Anakuta": *strong, wise man.*

While Eielson recovered from having a finger amputated, Wilkins continued exploring north of Point Barrow with Graham, but soon the season was over and they had to stop for the year. Islands upon which Wilkins had hoped to establish meteorological stations did not exist, but the team had collected much valuable information about the region. Wilkins was determined to come back next season and fly over the polar cap from Alaska to Spitsbergen.

OVER THE TOP OF THE WORLD!

The Lockheed Vega

After two years without hoped-for results, NANA and the Detroit backers pulled out, leaving Wilkins on his own. They paid off his debts and let him keep the two Fokkers.

In San Francisco to prepare for the continuation of his expedition, Wilkins was looking out of his hotel window one day when he spotted a streamlined monoplane racing through the air, the perfect plane for him and Eielson! He tracked it down to Oakland Airport—a Lockheed Vega, the first of its kind, made in Los Angeles especially for George Hearst, son of newspaper magnate William Randolph Hearst. Wilkins immediately ordered one built for himself. To pay for it, he sold the *Detroiter* to Australian aviator Charles Kingsford Smith.[23] Without big media sponsorship influencing his decisions, Wilkins decided to keep the new expedition small, though he retained the name "Detroit News-Wilkins Expedition" in gratitude for the newspaper's previous support.

Painted bright orange, the Vega had the same type of engine used by Byrd to fly to the Arctic and by Charles Lindbergh to cross the Atlantic. After many practice tests, the plane was shipped from Seattle to Seward to Fairbanks, arriving on February 26, 1928, accompanied by Wilkins and Eielson. On March 19 they flew to Point Barrow to take up where they had left off the year before.

By April 15 they were ready to go. With Eskimo help they had cleared a mile-long runway, and at 10:15 A.M. Wilkins and Eielson set off in a northeasterly direction from Barrow with just enough fuel to get them the 2,200 miles to Spitsbergen. Wilkins had chosen not to fly over the North Pole—the shortest and most direct route—as that would be duplicating the work done by Byrd and Ellsworth. Instead he charted a complex course across an area never flown over before, spanning ten time zones, where he hoped to discover land and take readings. Should the flight succeed, it would be the longest non-stop airplane flight in the Arctic.

Using the sextant to gauge their position, every hour Wilkins corrected their course, all the while sending and receiving wireless messages, shooting moving and still pictures, and looking for signs of land through two windows installed on the plane's floor. Through blizzards they flew over Greenland, and after twenty hours and twenty minutes they landed on a solid runway of ice, on what Wilkins calculated to be one of the islands in the Spitsbergen group (they later learned it was Dead Man's Island). Wilkins recalled the moment: "We must have sat there for some time, breathing heavily in our relief to be once more on terra firma. . . . We looked at each other and smiled our satisfaction as I made a record of the exact time."[24] Coming down in a severe storm, they waited inside the plane for five days until the weather cleared.

When they took off again they spotted the radio masts of Green Harbor, Spitsbergen, about thirty miles from where they had been stormbound. Here they were finally able to inform the rest of the world that they had succeeded.

Out of fuel, they were stranded in Green Harbor for three weeks until an ice-breaking steamer came and transported them and the plane to Tromsö in northern Norway, arriving May 15. From there they traveled to Bergen where they were officially feted by King Haakon for "the greatest effort in the history of aviation." Celebrations continued in other parts of Europe, where Wilkins was honored by geographical societies and scientific organizations.

Section 1

"All the News That's Fit to Print."

The New York Times.

THE WEATHER

VOL. LXXVII..No. 25,656. NEW YORK, SUNDAY, APRIL 22, 1928. FIVE CENTS

WILKINS FLIES FROM ALASKA TO SPITZBERGEN IN 20½ HOURS; COVERS POLAR SEAS IN TINY PLANE AND FINDS NO LAND; ISOLATED FIVE DAYS ON UNINHABITED ISLAND NEAR GOAL

SINCLAIR ACQUITTED OF OIL LEASE FRAUD; JURY OUT 2 HOURS

Verdict Causes Surprise in Capital and Stirs Bitter Comment by Senators.

"DISGUSTING," SAYS NYE

"Can't Convict Million Dollars," He Complains—Other Cases May Be Dropped.

JUROR EXPLAINS VERDICT

Rain, Sleet and Hail End The Menace of Forest Fires

NEW STATE BUILDING IS SUDDENLY HALTED

Quicksand and Muck Raise the Estimated Cost of Foundation From $244,900 to $1,000,000.

SITE A FILLED-IN POND

Contractors Bored 120 Feet to Reach Rock—20-Foot Piles Now Called Dangerous.

ROUTE OF THE WILKINS PLANE ACROSS POLAR REGION.

RELIEF TO BREMEN DELAYED BY ILLNESS OF PLANE'S PILOTS

Bennett and Balchen Indisposed at Lake Ste. Agnes—Schiller Likely to Go.

START FOR ISLAND TODAY

STEFANSSON JOINS BYRD IN ACCLAIMING FLIGHT OF WILKINS

Commander Rejoices in Conquest of Hitherto Unexplored Polar Region.

MAY SETTLE LAND QUESTION

TRIP ENDED BEFORE START

Wilkins and Eielson, an Alaskan Flier, Left Point Barrow Unannounced—Flew Route Below the Pole.

CALLED THE GREATEST FEAT OF ALL

Amundsen Hails the Flight as Far Exceeding Any Hitherto Accomplished—Success Crowns Wilkins on Third Attempt After Many Discouragements.

By CAPTAIN GEORGE H. WILKINS.

[BPCRC_W]

Roald Amundsen had warned him: "What you are trying to do is beyond the possibility of human endeavor."[25] After Wilkins' success, Amundsen said: "No flight has been made anywhere, at any time, which could be compared to it."[26] The American Geographical Society's chief surveyor commented: "[Wilkins] accomplished a feat in navigation which can be confidentially declared unparalleled in

Eielson and Wilkins arrive to a hero's welcome in Norway [BPCRC_W]

the history of flying."[27] Wilkins described the experience in his book, *Flying the Arctic,* published in 1928.

In Paris Wilkins had learned that he would be knighted, the ceremony to take place in London on June 14, 1928. The knighthood was not only for his heroic flight but also for his fifteen years of scientific and exploration work.

SUZANNE BENNETT

When Wilkins—now Sir Hubert Wilkins*—and Eielson returned to New York on July 2, 1928, they were greeted with a hero's welcome. The Australian consul had invited a fellow Australian, Broadway actress Suzanne Bennett, to be on hand for the festivities.

Born Suzanne Evans in Walhalla, Victoria, in 1893,** she was the daughter of engine driver John Evans and his wife Louisa Whitlow,

*He chose to use his middle name, Hubert, instead of George.

**Suzanne claimed to have been born in 1901, even in her passports. However, the license from her first marriage, which took place in 1915, states her age at the time as twenty-one.

Sir Hubert and Lady Suzanne Wilkins around the time of their marriage [MR]

and the oldest of four children. In 1915, at twenty-one, Suzanne married Oscar Bennett, a salesman and former military aviator; they divorced in 1922. By then she was performing on the Melbourne stage as a Tivoli showgirl.

In October 1922, Suzanne sailed from Sydney on the RMS *Niagara*, traveling to England via Hawaii and the United States. She listed her occupation as "singer." Visiting her aunt on a stopover in New York, she landed a part in a Broadway musical and decided to stay. She had roles in *The Gingham Girl, Innocent Eyes, The Rat*, and *Earl Carroll's Vanities*, before sailing to England on the SS *Aquitania* in December 1925. On the ship's record, she is listed as "actress" and the date of her birth as 1901. She returned to New York on the SS *Homeric* on January 13, 1926. For the next two years she performed on Broadway in *Port o' London*, *Nic Nax of 1926*, *Merry-Go-Round* and *What Do We Know?*

When Suzanne met Sir Hubert on July 2, 1928, she was playing in *The Cyclone Lover*. Almost at once the actress and the explorer became a pair, enjoying a few dinner dances before Sir Hubert, together with Ben Eielson, took off on a two-month lecture tour across the United States to raise money for their next expedition. During his absence, Suzanne starred in *Guns*, a Broadway production that ended its run right before Sir Hubert returned. He was back for only two days before sailing off again to South America, but before he left he proposed marriage to Suzanne and she accepted.

RACE TO THE SOUTH POLE

The purpose of Sir Hubert Wilkins' latest expedition was to explore the continent of Antarctica by air, something no one had done before. He would fly the same Lockheed Vega (now renamed the *Los Angeles)* that had carried him over the top of the world. William Randolph Hearst offered $25,000 in exchange for exclusive rights, plus a bonus if Wilkins made it to the South Pole. At the same time, Admiral Byrd, who had purportedly flown over the North Pole on May 9, 1926, was preparing to fly over the South Pole, and the Hearst papers promoted the event as a race to the South Pole between the two explorers. Despite the sensational headlines, Wilkins was not motivated by a contest between himself and Byrd; his object, as always, was to find land suitable for weather stations.

In contrast to Byrd's 100-man operation, Wilkins led a compact team of six, including pilots Ben Eielson and Joe Crosson, a wireless operator and two mechanics. He also purchased a second Lockheed Vega and named it the *San Francisco*.

On September 22, 1928, Wilkins and his group, together with the planes, sailed on the SS *Southern Cross* from New York to Montevideo, arriving on October 10. Here the expedition joined the SS *Hektoria* for the voyage south. En route Sir Hubert received a cable from a representative of the British government authorizing him to

The former luxury liner SS Hektoria, *which the team used as a base* [BPCRC_W]

claim new lands for Britain. On November 7, after a rough journey through thick pack-ice, the ship dropped anchor in the volcano-crater harbor of Deception Island.

The *Hektoria* was a former White Star liner that had been converted into a whale processing vessel and mother ship to five small whale catchers. Its luxury cabins remained intact, and Wilkins had arranged for his team to remain on board, using the ship as their base, from November to the end of February. This eliminated the need for bringing the usual heavy gear and equipment.

To their dismay, it was a warmer summer than anticipated. The harbor ice, intended for take-offs and landings, was too thin for a runway so they had to convert the planes to seacraft. They encountered another unexpected hazard in the form of a large population of albatross-like birds which flew along during takeoffs and often into the propellers.

As a last resort they decided to create their own runway on nearby hilly land covered in boulders and lava. Using picks and shovels, and with the help of whalers, the job took weeks. Several test runs were made, one resulting in near disaster when Eielson landed on too-thin ice and the plane began sinking nose-down in water. It took all available hands eighteen hours to drag it back out onto solid ice. While the salt-water damage was being repaired, Wilkins borrowed a catcher and explored the nearby islands, looking for potential landing strips. Finding none, they decided to extend the existing strip to half a mile.

With their goal being to fly over the Antarctic continent, Wilkins and Eielson lifted off at 8:30 a.m. on December 20, loaded with survival gear, a two-month supply of food, and enough fuel for 1,300 miles. On board was a short-wave radio set that could keep the wire operator advised of their position and establish contact between any

Exploring from the air [BPCRC_W]

Wilkins and Eielson returning from the first flight over Antarctica [BPCRC_W]

two points on the planet. With Wilkins sketching the terrain and shoreline features, mapping and charting mountain summits and using the sun to plot their position, they flew to Graham Land, covering in twenty minutes what had taken them three months on foot eight years earlier. Wilkins wrote: "This time I had a tremendous sensation of power and freedom—I felt liberated. . . . The contrast was most striking between the speed and ease of flying in by plane and the slow, blind struggles of our work along the coast a few years before. . . . No one had climbed to the Graham Land plateau summit, and I was thrilled to realize that for the first time, human eyes—our eyes—were going to see it."[28]

From there to the eastern coast of the Weddell Sea, Wilkins filmed and sketched landmarks of a heretofore uncharted region. What he believed to be a body of water separating Graham Land from the mainland he named "Stefansson Strait." After flying over what appeared to be island after island, they reached a vast plateau, which Wilkins named "Hearst Land."[29]

When they had used half their fuel they turned around, but not before Wilkins dropped a Union Jack onto the land below from the

Loading the Vega onto the ship [BPCRC_W]

floor hatch, claiming the land for Great Britain. Heading straight into a storm, they returned to Deception Island in time to dress for dinner at eight on the *Hektoria*, having covered close to 1,300 miles, most of it over unknown territory.

With the season coming to an end, they dismantled the planes and stored them in the whalers' warehouses with plans to return the following summer. In January they sailed back to New York via the Falklands, the Straits of Magellan, the west coast of South America and the Panama Canal.

In late January 1929, an editorial in *The New York Times* heralded Wilkins as a man who had "changed the face of the known earth."[30] He had proved the feasibility of exploration from the air as well as its shortcomings: As a land expedition later proved by drilling down, what Wilkins had believed to be straits and channels were actually fjords sitting on land above sea level, proving that Graham Land was not an island after all. But this discovery did not faze Wilkins; as a scientist, he was interested in facts not fame.

While Wilkins had indeed been the first to fly over Antarctica, the following year, on November 28, Byrd's team with senior pilot Bernt Balchen at the controls was the first to fly over the South Pole, thus winning the "race."

GRAF ZEPPELIN

Upon his return to New York in March 1929, Wilkins took William Randolph Hearst up on his offer to join the airship *Graf Zeppelin* on its inaugural flight around the world. The vessel—774 feet long, nine stories high, with five 50-horsepower engines—was the brainchild of the late Ferdinand Graf von Zeppelin. It was completed by Dr. Hugo Eckener, who had taken it on its initial voyage—from Friedrichshafen, Germany, across the Atlantic—in the fall of 1928. As the major sponsor of the globe-circling effort, Hearst had made one demand: the voyage had to begin and end in New York. Among the correspondents on board were Karl von Wiegland to cover the main news, Lady Grace Drummond-Hay to report on the human side, and Sir Hubert to cover the technical aspects of the flight.

On May 16, 1929, with Eckener again at the controls, the *Zeppelin* lifted off from Friedrichshafen, headed for Lakehurst Airfield just south of New York. After floating gloriously for several hours over Swiss lakes and French towns and valleys, alarming noises were heard. One by one the engines began cutting out, but Eckener mas-

Sir Hubert, von Wiegland, Dr. Eckener, and Lady Drummond-Hay [BPCRC_W]

View from the air with the shadow of the Graf Zeppelin [BPCRC_W]

terfully guided the vessel to an empty hangar in a nearby airship field south of Toulon, France. While the other passengers were placed under house arrest by French police suspicious of a German ship landing in their midst, the reporters, including Wilkins, managed to steal away to telegraph the news. Engines repaired, they glided back to Friedrichshafen where it took another six weeks to prepare the vessel for a second attempt to fly around the world.

Meanwhile, word came to Sir Hubert that Suzanne had been hospitalized with rheumatic fever, so while waiting he made a quick trip to New York to take care of her before rejoining the other reporters in Friedrichshafen.

On July 31, 1929, the *Graf Zeppelin* with several passengers and a crew of forty-one took off again for Lakehurst Airfield. There she took on an additional twenty passengers—an international group consisting mainly of journalists sponsored by news agencies. On August 7 the ship began its epic flight, gliding serenely for four thousand miles to Friedrichshafen, where it refueled. From there it went to Japan via Danzig, Lithuania and the Soviet Union. Besides filming, Wilkins spent much of the voyage studying Eckener's expertise in navigating the ship through the winds and weather. On August 19 they arrived in Tokyo for celebrations that included tea at the emperor's palace. While there, Wilkins made connections with bureaucrats and toured the Japanese aircraft industry, which he declared to be as up-to-date as its western counterparts.

Four days later the *Zeppelin* lifted up again and set a record for crossing the Pacific Ocean in seventy-nine hours, reaching San Francisco and then traveling south down the dark evening coast to the sprawling, lit-up city of Los Angeles. Here they landed and were escorted through throngs of well-wishers to the Ambassador Hotel where Hearst was waiting with a grand welcome gala attended by Hollywood notables. Both Eckener and Wilkins gave speeches. After the brief stopover, passengers and crew returned on board at midnight for the last leg of the journey back to New York. After a near-disastrous take-off in the hot, windless air, in which all excess weight—including the entire water supply—was thrown overboard as an emergency measure to help propel the craft over dangerous power lines, the passengers settled down to cross the country. Wilkins wrote, ". . . Still thirsting under Prohibition Law, we drank champagne in lofty disdain of water and the law. We washed ourselves in eau de cologne!"[31]

After thirteen days' flying, covering a record-breaking 20,000 miles, the *Graf Zeppelin* touched down at Lakehurst Airfield on August 29. While Eckener was whisked off to meet President Herbert Hoover, Wilkins avoided the publicity and parades and rushed into Suzanne's waiting arms; immediately they took off together for Cleveland where they were married the next day.

SECOND ANTARCTIC EXPEDITION

By this time plans were finalized for the next Antarctic expedition, to be financed by leftover Hearst funds and by the British who wanted to stake claim on newly discovered land. Wilkins took on two new pilots,[32] Parker "Shorty" Cramer and Alward "Al" Cheesman.

The Vega on skis [BPCRC_W]

Leaving his new bride behind in New York to begin her life as "the loneliest wife in the world,"[33] Sir Hubert and his team boarded the *Hektoria*, arriving at Deception Island on November 1929, where their two stored planes and other equipment were waiting. To their dismay, they were unable to establish a base as the icepack had receded six hundred miles since the previous year.

The British Colonial Office had offered the use of its research vessel, SS *William Scoresby*, should the team need to travel farther south. Wilkins took them up on the offer, and while waiting for the ship, the team made short flights charting nearby land. When the *Scoresby* arrived in mid-December, they loaded the Vega and equipment on board and sailed to the whaling harbor of Port Lockroy, on the west side of Graham Land. On December 19, they took off from the water using floats and explored the coast for landing sites. Having located a possibly suitable bay, the *Scoresby* sailed south in search of it. Along the way they encountered thousands upon thousands of giant Emperor penguins who stood watching them in eerie silence. At their destination, the Vega was lowered onto the ice with skis attached, but the ice began to melt and they hastily hauled the plane back on board. A sudden storm followed which lasted three days.

Had they taken off, the chances of landing safely would have been nil. Again the hand of fate had stepped in to rescue them.

When the weather cleared, they navigated south to a spot north of Charcot Land where they dropped anchor and found suitable spots to take off and land for a round of exploring before being driven back by a violent storm. Flying over Charcot Land, Wilkins determined it was an island, not part of the mainland, and claimed the land for Britain.*

The *Scoresby* continued westwards into the Bellingshausen Sea and the three made some progress exploring the region. In the course of charting and mapping Wilkins had located many possible sites for future weather stations, but in January he called an end to the expedition citing bad weather and difficulty in finding places to take off. Back in Buenos Aires, he sold the two planes to the Argentine government and by March he was back on American soil.

HONEYMOON

Suzanne, Sir Hubert and Lincoln Ellsworth at Schloss Lenzburg [BPCRC_W]

*The body of water separating it from the mainland is known today as Wilkins Sound.

Taking time out between expeditions for a belated honeymoon, Sir Hubert and Lady Wilkins boarded the *Graf Zeppelin* on June 2, 1930, and flew to Germany. On June 3, Suzanne telegraphed a message from the sky to newspapers around the world: "We are sailing on a magic ship betwixt heaven and earth."[34]

In Friedrichshafen they were met by Lincoln Ellsworth, the explorer and financier who had flown over the North Pole with Roald Amundsen in 1926. Ellsworth invited the couple to stay at his Swiss castle, Schloss Lenzburg. For a month the newlyweds enjoyed walks and visited tourist sites.

THE *NAUTILUS*

In conversations with Ellsworth at Schloss Lenzburg, Wilkins shared his plans for a revolutionary mission—exploring the underside of the polar ice cap in a submarine. Ellsworth agreed to help finance the venture. As a fundraising tool, Wilkins published a book in advance, *Under the North Pole*, outlining his plans. Most believed the idea to be impractical, if not impossible, but Wilkins prevailed and arranged for the U.S. Navy to lease him a scrapped submarine for $1 a year.

Modified and outfitted for Arctic use and renamed the *Nautilus*, the vessel was launched in Brooklyn, New York, on March 24, 1931. Suzanne christened it with a bucket of ice cubes instead of champagne to accommodate Prohibition laws. On hand was Jean Jules Verne, grandson of Jules Verne, the author of *Twenty Thousand Leagues Under the Sea*, the *Nautilus* having been named after the fictional submarine in the book. Hearst and Paramount Pictures contributed to the financing in exchange for exclusive newsreel footage.

On June 4, 1931, with a crew that included a commander, an oceanographer, a meteorologist and polar scientists, the Wilkins-Ellsworth Trans-Arctic Expedition set off from Cape Cod, Massachusetts, to Plymouth, England. From there they planned to go to Norway and dive beneath the Pole to Alaska. Trouble began on June 13, a thousand miles out at sea, when they experienced major mechanical trouble during a violent storm. Unable to restart the engines

The Nautilus *and crew* [BPCRC_W]

and faced with life-threatening conditions on board, the captain sent out an SOS. Rescued by the American battleship *Wyoming*, the *Nautilus* was towed to Ireland for repairs. Back at sea, the vessel suffered more breakdowns and was in dry dock in Plymouth on July 15 when Sir Hubert gave Edward, the Prince of Wales, a tour of inspection.

From there the *Nautilus* plowed on to Bergen, Norway, then to Spitsbergen where final maintenance took place before they ventured forth into the Arctic ice. Finding a spot about six hundred miles from the North Pole to begin the experiment, it was discovered that the submarine would not submerge. A diver sent down to investigate reported that both diving rudders had disappeared. This meant that the vessel could not be properly controlled nor rise quickly once underwater. Sabotage? Could they go on? Wilkins insisted that although they were badly handicapped they should not give up.

On August 31, after several tests, the crippled *Nautilus* finally managed to go under, "trembling as if subjected to a giant earthquake,"[35] to a depth of thirty-seven feet, the captain not daring to go further down. When the engines were turned off, the men gazed in awe at a sight never before witnessed by human eyes: the underside of an ice floe. As the light filtered in, an array of colors danced about—"minarets, domes, majestic cathedrals, geometric blocks . . . in the crystal-clear blue water."[36]

Several other brief dives were made and filmed, some marine-life samples gathered, but without a properly equipped vessel Wilkins was forced to call an end to the mission. Hearst agreed and ordered the crew to return, citing their welfare.

Back in Spitsbergen by the first week of September, Wilkins debarked the *Nautilus* and traveled to London where Suzanne was waiting to sail back with him to New York. The submarine continued on to Bergen where it was scuttled[37] and the crew was sent home on a Norwegian steamer. Hearst was displeased with the venture and refused to pay, as did Paramount Pictures. Lincoln Ellsworth, on the other hand, forgave Wilkins the $20,000 debt that was owed him.

THE ELLSWORTH TRANS-ANTARCTIC EXPEDITION

The Polar Star [BPCRC_W]

Wilkins spent most of 1932 lecturing to get out of the debt he incurred by the *Nautilus* misfortune. To show his gratitude to Ellsworth, he agreed to help him realize his dream of being the first man to fly across the Antarctic continent. In early 1933, while Ellsworth (who had married for the first time at age fifty-three) was honeymooning in the Pacific, Wilkins organized the first of four Antarctic expeditions for him. In California he located a unique Northrop seaplane, the *Polar Star*, which could take off from water at shorter

The Wyatt Earp [BPCRC_W]

distance and land at slower speeds than usual; in Norway he found a ship (renamed the *Wyatt Earp*) whose semi-diesel engine gave her extraordinary power to break through ice floes. He recruited Bernt Balchen, who had flown with Admiral Byrd over the South Pole in 1930, as senior pilot and hired a captain and crew with Antarctic whaling experience. Wilkins himself would serve as second-in-command, press reporter, and captain of the base.

On July 29, 1933, the team sailed from Bergen to Antarctica via the west coast of Africa, picking up supplies in Dunedin in New Zealand and arriving at Little America the first week of January 1934. The *Polar Star* was reassembled and several test flights were performed, but on January 12 misfortune struck when the runway upon which it was parked cracked open and the plane sank partially into the water. It was pulled out but was so damaged that the expedition was called off for the year. The plane was shipped back to Northrop in California for repairs.

The Wyatt Earp *trapped in the ice* [BPCRC_W]

Throughout most of the winter, Sir Hubert remained in Dunedin with the dry-docked *Wyatt Earp* to supervise improvements. He also gave lectures throughout New Zealand. With time to spare, he accepted a June speaking engagement in London, where he rendezvoused with Suzanne for a few weeks before returning to Dunedin via Italy, Egypt, Palestine, Iraq, India, Thailand, Burma, Indonesia and Australia, rejoining the *Wyatt Earp* in early August.

SECOND ELLSWORTH ANTARCTIC MISSION

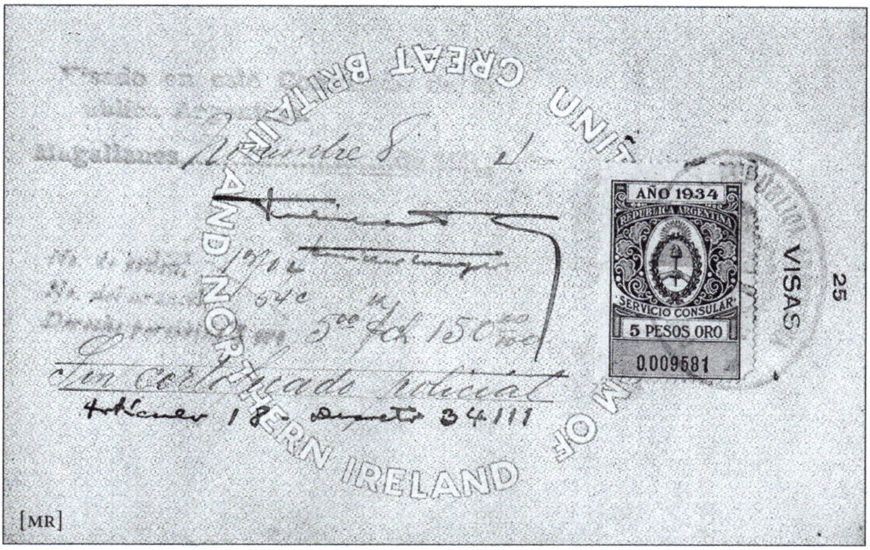

[MR]

When the team returned to Dunedin, they found the boat in good shape and the plane, now carrying floats as well as skis, thoroughly restored. This time they decided to set out from the opposite end of Antarctica, from Wilkins' old haunt, Deception Island. On September 19, 1934, the *Wyatt Earp* cast off for the 4,000-mile journey. When they arrived, bad weather gave them many setbacks, and on the eve of their planned take-off, a connecting rod broke when the mechanic started the engine. The flight was postponed while replacement parts were shipped to Magallanes, a port six hundred miles away in southernmost South America, the *Wyatt Earp* sailing the round trip to collect them.

On December 30, Ellsworth and Balchen made another start at being the first to fly across the Antarctic continent. Within a few hours they ran into such severe weather that Balchen, against Ellsworth's wishes, turned back fearing it would be suicidal to continue. Ellsworth was furious and refused to work with Balchen again. From then on weather conditions grew worse and they were compelled once more to give up for the year. The *Polar Star* was dismantled and stowed for the journey back to Montevideo, where the *Wyatt Earp* was laid up to resume operations the following spring.

THIRD ELLSWORTH ANTARCTIC MISSION

Lincoln Ellsworth and Sir Hubert Wilkins [BPCRC_W]

Between expeditions Sir Hubert joined Suzanne in New York and they traveled to Europe in June before he returned to Montevideo in early August. He tried to persuade Ellsworth to give up his Antarctic mission, but reluctantly agreed to organize a third attempt. This time he hired two new pilots, Herbert Hollick-Kenyon and J. H. Lymburner. Early in November 1935, the team boarded the *Wyatt Earp* and sailed to Dundee Island, 160 miles east of Deception Island, where they established a base and found good conditions for flying.

On November 23, leaving Wilkins behind to oversee base operations, Ellsworth and Hollick-Kenyon took off, planning to take fourteen hours to reach the Ross Ice Shelf and claim victory. Instead, the trip took twenty-two days and covered about 3,500 miles, with forced landings and delays due to bad weather. Radio communications were disabled, and when nothing was heard, Wilkins—by prior arrangement with Ellsworth—began the 5,000-mile journey of bringing the *Wyatt Earp* around to the Little America side to look for them.

Ellsworth and Hollick-Kenyon, meanwhile, finally reached their destination on December 15, twelve miles from Little America, where they abandoned the plane and trekked to Byrd's old camp on foot. Ellsworth was triumphant, his great Antarctic dream accomplished. The duo, who were reported missing, were eventually picked up by an Australian ship which hooked them back up with the *Wyatt Earp* on January 20. The *Polar Star* was recovered and secured to the ship's deck, and the party sailed back to New York, arriving on April 19, 1936. Ellsworth donated the historic plane to the Washington Museum.

LADY WILKINS ADOPTS A DAUGHTER

FAMILY GROWS

WHILE HER EXPLORER-HUSBAND passed Christmas in the polar regions, Lady Hubert Wilkins announced adoption of a 6-year-old girl, whom she has re-christened Suzanne, after herself. Sir Hubert will be appraised of the adoption over the short wave radio tomorrow.

[MR]

It was during this third Antarctic expedition that Lady Suzanne Wilkins decided to adopt a six-year-old girl, unbeknownst to her husband. The press reported that Lady Wilkins broke the news to Sir Hubert in Antarctica via short-wave radio on New Year's Day 1936 from the Ritz-Carlton Hotel in New York. The same item stated that Suzanne named the girl Suzanne after herself, had known her since infancy, and planned to send her to the same convent school Suzanne claimed to have once attended in England. It said that the adoption was not finalized, that a probationary period would have to elapse. Though newspaper clippings about the planned adoption were found pasted into one of Suzanne's personal photo albums, the story ends here. Nothing more has been uncovered about the mysterious girl and what happened to her, or how Sir Hubert had reacted to Suzanne's plans.

THE *HINDENBURG* VOYAGE

No sooner had Wilkins returned from his latest Antarctic venture than he was off again, this time with Suzanne, for Germany. They were joining Dr. Eckener on the airship *Hindenburg*, on her first transatlantic voyage, from Friedrichshafen to Lakehurst, New Jersey.[38]

Lifting off on May 6, 1936, the crossing was without incident but for one outstanding event—as the dirigible floated over New York, Suzanne was invited to sing. She chose a popular song of the day, "I'm in the Mood for Love," and her performance was broadcast over New York radio.

Sir Hubert and Lady Wilkins being interviewed after the voyage [MR]

THE SEARCH FOR LEVANEVSKY

Refueling the Russian-built plane during the search for Levanevsky [BPCRC_W]

Wilkins was still hoping for another submarine mission when, in August 1937, he was contacted by a representative of the Soviet Embassy, who wanted to consult with him concerning the disappearance of aviator Sigismund Levanevsky, known as "Russia's Lindbergh." Levanevsky and five others had taken off from Moscow on August 13, crossed the North Pole and were headed for Fairbanks when they radioed trouble and were not heard from again.

Wilkins was asked to organize a search to find the missing men. He located a plane and formed a team consisting of pilots Al Cheesman and Herbert Hollick-Kenyon, engineer Gerald Brown, and U.S. Navy radio operator Ray Booth. On August 20 they headed north to Coppermine where they established a base. For a month they crisscrossed the area between Siberia and Northwest Alaska, flying 13,000 miles, but found no signs of the missing pilot and his crew. By late September, with winter and freezes coming, they abandoned the search and returned to New York.

Not wanting to give up, the Soviet Embassy provided Wilkins with a new plane that was better equipped to withstand Arctic winter conditions, and the search was resumed late in November 1937. Allen Dyne replaced Booth and Brown and a base was set up at Aklavik. For months they flew over the region, and though they were unable to find Levanevsky, the flights proved the feasibility of standard aircraft for Arctic operations. On March 25 the search was called off and shortly thereafter Sir Hubert traveled to Moscow to be presented with official thanks for his attempt to find the Soviet heroes.

EXPERIMENT IN ESP

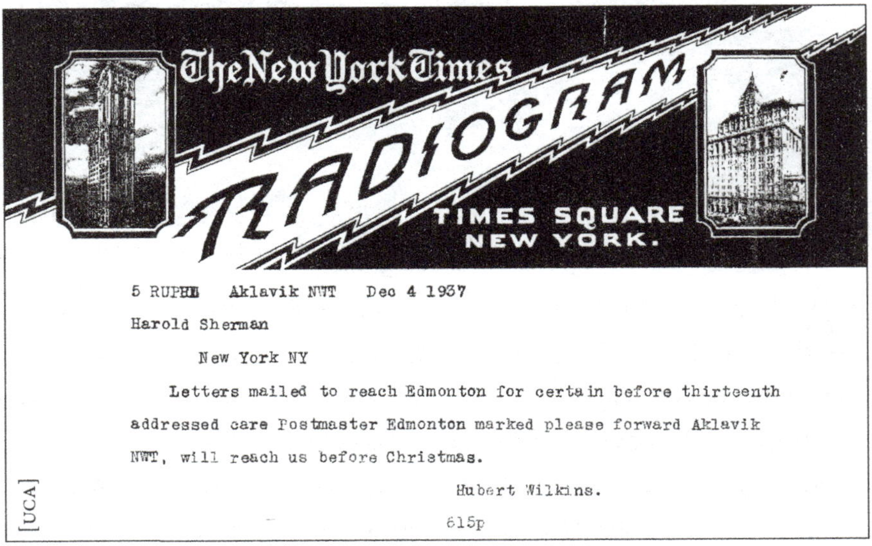

The New York Times
RADIOGRAM
TIMES SQUARE
NEW YORK.

5 RUPHE Aklavik NWT Dec 4 1937

Harold Sherman

New York NY

Letters mailed to reach Edmonton for certain before thirteenth addressed care Postmaster Edmonton marked please forward Aklavik NWT, will reach us before Christmas.

Hubert Wilkins.

615p

[UCA]

It was during the second attempt to find Levanevsky that Sir Hubert and author Harold Sherman—a New York friend and fellow City Club* member—conducted a pioneering experiment in mental telepathy.

Both men believed in extra-sensory perception and worked out a test: Three evenings a week, at an agreed-upon time, Sir Hubert, over three thousand miles away in the Far North, "sent" his thoughts to

---

*The City Club of New York was founded in 1892 to promote effective and honest government in New York City. The clubhouse was located at 55 W. 44th St. Guests (such as Sherman) stayed on the sixth floor.

Sherman, who "received" them in his study in New York. Sherman typed up his impressions, sealed them in an envelope, and mailed them to Dr. Gardner Murphy, professor of psychology at Columbia University. When Sherman's impressions were compared with Wilkins' diary entries, both men were astounded at the accuracy of so many of Sherman's visions. Their jointly-written book, *Thoughts Through Space,* published in 1942 by Creative Age Press, gave a full account of their experimentation.

LAST ANTARCTIC EXPEDITION WITH ELLSWORTH

Wilkins would have preferred another submarine mission to the North Pole, but with no funding forthcoming, he agreed instead to a fourth expedition to Antarctica with Ellsworth, this time to fly over the Antarctic Circle. The *Wyatt Earp,* having been refitted in Norway, was brought back into service, along with two planes, pilots Herbert Hollick-Kenyon and J. H. Lymburner, and an all-Norwegian crew. On August 16, 1938, the vessel departed New York for Antarctica. On the way they took on supplies in Recife, Brazil; picked up Ellsworth (who had been on a safari) in Cape Town, South Africa; and stopped off at the Kerguelen Islands, in the South Indian Ocean, for maintenance. From there they set off on November 17 for a stormy forty-five-day passage to the Antarctic mainland.

Unbeknownst to Wilkins, Ellsworth had a secret agenda. There was still uncertainty as to who owned Antarctica, with no nation recognizing the others' claims. Ellsworth had received top-secret instructions from the U.S. State Department to stake claim to any unexplored areas, even if others had already claimed them. One of these areas was on the Davis Sea coast of Eastern Antarctica, from where Ellsworth intended to launch his flight. This was regarded as Australian territory, having been claimed by Wilkins' rival, Sir Douglas Mawson. The claim was disputed because Mawson had not physically set foot on the land at the time. Ellsworth could not keep his secret long and on January 7 he confessed to Wilkins the real purpose of the undertaking. Wilkins was now in a moral predicament: Should he be loyal to his own country or fulfill his obligation as an employee of an American expedition?

Sir Hubert solved the dilemma by going ashore on the pretext of looking for geologic specimens. With pilot Lymburner as witness, he raised the Australian flag; the next day, January 9, 1939, he repeated the ritual on another Australian territory, leaving evidence of

The Australian flag left behind by Wilkins [BPCRC_W]

his visit in the form of a copy of *Walkabout* magazine, which was discovered in 1957 by a party of Australian scientists. Ellsworth, meanwhile, had flown over the area dropping the American flag. Wilkins alerted Australian officials suggesting they establish a base there to refute Ellsworth's claims; he even offered to set it up himself using Ellsworth's boat and equipment. The Australian government was willing to negotiate a deal, and even purchased the *Wyatt Earp*, but the project was cancelled when the Second World War came along.

THE FARM AND WINSTON ROSS

Suzanne with Winston Ross on the porch at Walhalla [MR]

In the summer of 1939 Sir Hubert returned to New York, where he learned that he had become a landowner for the first time in his life, Suzanne having purchased a farm in Pennsylvania. Wilkins enjoyed having a home in the country, and spent time between missions fixing it up and putting in a pond. The Wilkinses named the property Walhalla, after Suzanne's birthplace. There he housed his vast collection of writings, artifacts, photographs, films, and other memorabilia from his many travels.

Around the same time, Sir Hubert and Lady Wilkins met Winston Ross, a singer and actor, at a cocktail party held at the Dutch Consulate in New York City.[39] Ross was the hired entertainment, and after Wilkins complimented him on his singing, an enduring friendship/working partnership began among the three. Ross was to become the Wilkinses' secretary/assistant and Suzanne's escort when Wilkins was away on his travels. The true nature of the close relationship between Suzanne and Winston Ross remains an open question.

OUTBREAK OF WORLD WAR II

When war broke out again, Wilkins offered to serve in combat for both the Australian and British military but was politely rejected by both as being too old. Looking for a way to support Britain in the war, and well aware that Britain's great weakness was a shortage of aircraft, Wilkins approached the Detroit auto and aircraft makers to see if they could help. The manufacturers united to form the Aero Products Division and asked Wilkins to travel to England as their representative. He would negotiate with Lord Beaverbrook, Cabinet Minister for aircraft production and an acquaintance of Wilkins' since World War I.

Because of the war, Wilkins had to reach England in a roundabout way. After flying from Baltimore to Lisbon, he took the train to Paris just days before France fell to the Germans on June 14, 1940. He first tried to escape on a French transport aircraft heading for Britain, but it was shot down in a field. Finally, after passing through German lines by bicycle and on foot, he reached Nantes where he hitched a ride to England with the Royal Air Force. In London he met with Lord Beaverbrook, who was enthusiastic about receiving help from Detroit until Wilkins told him it would take at least three months for planes to begin arriving. "Three months . . . and we hardly know if we shall still be here in three weeks," said Beaverbrook,[40] referring to a seemingly impending invasion by the Nazis. However, Lord Beaverbrook arranged for Wilkins to meet with experts and work out a deal for Detroit to furnish planes for the British war effort.

ESPIONAGE IN ASIA

Upon his return to the United States, Sir Hubert went into top-secret work for the Secret Intelligence Office of Strategic Services,* covered by his work for the aviation industry. Even Suzanne, supposedly, did not know what his work involved; when she asked him about it, he replied, "Can you keep a secret? Well, so can I."

*This U.S. agency was formed during World War II and was a predecessor of the Central Intelligence Agency (CIA).

Wilkins in Asia [BPCRC_W]

On June 2, 1941, after spending a few days in Los Angeles visiting friends (including Harold Sherman who was there on a screenwriting assignment), Wilkins boarded a Japanese steamer headed for Hong Kong on a special mission for the U.S. government. He went through Japan, Manchukuo,* and the western Chinese provinces which Chiang Kai-Shek had managed to protect from Japanese invasion. Sir Hubert claimed to have met with the leader, whom he described as "just another Chinese warlord."[41]

*Manchukuo was a puppet state created by Japan in 1932 out of the three historic provinces of Manchuria. It was dissolved in 1945.

From Kunming in southern China he journeyed over the Burma Road and along the Irrawady River to Thailand, and from there overland to Singapore. At the Raffles Hotel he dined with the Japanese consul, who drunkenly revealed to Sir Hubert that plans were in motion to bomb Pearl Harbor in four months' time. When Wilkins asked why he was telling him that, the consul replied that he knew Wilkins was a spy for the Americans and that nobody would believe him when he told the story.[42] From Malaya he flew to Batavia in the Dutch East Indies, returning to the United States via Australia in late October.

URANTIA

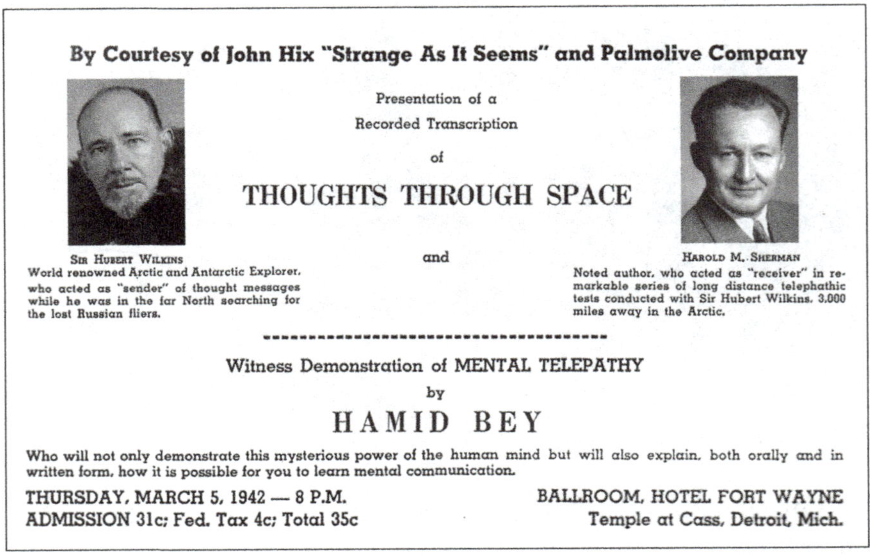

[UCA]

At some point during this period Harold Sherman told Wilkins about a new "revelation" he had been exposed to at the residence of a Dr. William S. Sadler in Chicago. The revelation consisted of thousands of typewritten pages of metaphysical teachings channeled from higher realms. The manuscript could be read only on the premises by members of a specially-formed Forum who had signed oaths of secrecy. It was intended to be published eventually as a book. Wilkins was intrigued.

The Wilkins-Sherman collaboration, *Thoughts Through Space*, was published on February 20, 1942. On that day the two men embarked on a several-weeks-long book tour in the East and Midwest. Between engagements, they visited Chicago and Sherman introduced Sir Hubert to Dr. Sadler, who admitted him to the Forum on March 5, 1942. From that day until the material was published as *The Urantia Book* in October 1955, Wilkins returned to Chicago at every opportunity to read and study the papers.*

THE LATER YEARS

Sir Hubert Wilkins never gave up his Australian nationality, but for the rest of his life he continued to work for the United States government, which was pleased to make use of the explorer's unique knowledge and experience. In June 1942, Wilkins was hired by the product development branch of the Army's Quartermaster General, working under General Georges Doriot. This department was concerned with the designing and provision of military equipment for use in arctic, desert, jungle, mountain and tropical environments. Wilkins spent part of the month in Washington, D.C., where (as he told Harold Sherman) "for the first time in my life I have a swivel chair,"[43] and the rest of the time traveling—interviewing, inspecting, and writing reports. Within the first month he had redesigned flea powder, a dog harness, a sledge, snowshoes, a tropical helmet, and a motor snow sled.

After leaving the Quartermaster Office in early 1946, Wilkins spent a term lecturing at Swarthmore College. Thereafter he was offered a long-desired position with the U.S. Weather Bureau, "putting together five weather outfits which we expect to install this summer. The job is confidential—the Arctic plans, that is, not my appointment as 'Advisor to the Arctic Administrative Staff, U.S. Weather Bureau, Arctic Projects.' The Navy and Army Air Force are working with us and I expect to go north with at least five ships in July."[44]

*More details on this aspect of Sir Hubert's life are found in the following section, "Wilkins the Truth Seeker."

In the Arctic as a consultant in later years [BPCRC_W]

By October 1946, Wilkins was back with the Quartermaster General but "in a higher echelon—office of Chief of Staff now."[45] In December he joined the Navy as consultant in the Naval Research Section. In addition, he was employed by the Arctic Institute of North America, headquartered in Montreal, lecturing on subjects ranging from arctic clothing and equipment to penguins in the Antarctic. "All of these jobs," he wrote to Sherman, "will keep my fingers out of my pockets most of the time."[46]

*Lowell Thomas was an American writer, motion picture producer and broadcaster. He was a great friend of Sir Hubert Wilkins and in 1961 published *Sir Hubert Wilkins, His World of Adventure.*

His duties took Wilkins from his base in Washington, D.C., to faraway places for many months at a time—to the Arctic, Antarctica, the deserts of Arizona and the Middle East, and the jungles of Panama. Throughout the years Wilkins never stopped lecturing on polar exploration—"getting more offers for lectures than I can take"[47]—crisscrossing the United States, Canada and Alaska. Though he lived mostly in hotels, trains, buses, ships and planes, Sir Hubert managed occasionally to spend time with Suzanne, either at their apartment in New York or on their farm in Pennsylvania.

In September 1952 he wrote to Sherman that he had just returned from a trip through Pakistan, Iraq, Syria, Turkey, Lebanon, Egypt, Libya, Algeria, Morocco, France and England, "not just a 'flying' trip but had to visit, all told, 65 cities and make reports about them. Took me about 75 days and that meant *some* travel."[48] The purpose of the journey was to study the colors of the major deserts of the world, in order to produce the most appropriate military camouflage clothing. His task involved taking color photographs at various times of day and collecting sand samples.

By 1957—the International Geophysical Year (IGY)—Wilkins' dream of establishing weather stations in Antarctica had become reality: Twelve nations, including the U.S.A. and the U.S.S.R., joined together in peace to establish fifty weather stations in the Antarctic.

In September of that year he joined a party that included his former pilot Bernt Balchen, Lowell Thomas Sr.* and Jr., and Donald MacMillan* to fly once more across the Arctic. Starting from Point Barrow, they flew north and visited IGY scientists on an ice floe in the Arctic Ocean, then continued on over the North Pole to Greenland.

Less than two weeks later Wilkins was back in Antarctica, visiting many of the IGY stations and the old Scott and Shackleton bases. In June 1958, he wrote Sherman that he had just finished a tome-sized report on his Antarctic trip—"a damning indictment of the Navy—poor seamanship, poor management, ignorance, indifference except

---

*Donald MacMillan was an American explorer, sailor, researcher and lecturer who made over thirty expeditions to the Arctic.

Clockwise from top left: MacMillan, Balchen, Wilkins, Thomas [BPCRC_W]

in the strictly-adhered-to practice of carrying on management from an alcoholic plateau. I was utterly disgusted and disappointed and hope that my frank reporting . . . will aid in leading toward greater success in their Polar endeavors."[49]

In the same letter, Wilkins mentioned that he was writing a script for an Army film to be called "Protection for the Soldier on the Battlefield," illustrating what his department had done to protect man from all environments. He said that he hoped to film it in the Arctic and at Mount Washington Observatory in New Hampshire the following winter.

"One of these days," he had written to Sherman in 1956, "I will probably retire or get fired from the Army and then I will have time to think about private affairs."[50] But before either could happen, on December 1, 1958, Wilkins died of a heart attack in his hotel room in Framingham, Massachusetts, where he stayed when working at the Quartermaster Research laboratory in nearby Natick. He was seventy. He had evidently been working on his car and was still wearing his overcoat when his body was found by a chambermaid. According to Suzanne, mysterious circumstances and unanswered questions surrounded his death. She confided her suspicions to Sherman, saying, "I shall always wonder."[51] She arranged to have her husband's ashes transported to the North Pole aboard the submarine USS *Skate* where they were scattered on the ice on March 17, 1959.

March 17, 1959: Wilkins' ashes are scattered at the North Pole [BPCRC_W]

WILKINS
THE TRUTH SEEKER

THE TRUTH SEEKER

My Father, I beseech support in my desire to worship,
To enjoy privilege without abuse,
To have liberty without license,
To have power and refuse to use it for self-aggrandizement,
So that the experience of living will lead me and my fellows
To greater spiritual reality.*

Sir Hubert Wilkins was a man of amazing versatility—photographer, filmmaker, war correspondent, aviator, naturalist, geographer, meteorologist, submariner, spy, author, lecturer, explorer of the North and South Poles and many points in between. Not only was he an explorer of facts about our world, he was also keenly interested in exploring the powers of the mind and trying to penetrate the mysteries of the universe.

Like many men of his social background and generation, he was brought up in a conventional church, but he later sought insight from beyond the confines of traditional Christianity. Eventually he encountered the Urantia text, and through his early embrace of it he bears the distinction of being the first internationally known person to accept and evangelize the Urantia Book as a major revelation of truth to this planet.

---

*Suzanne Wilkins described this piece of writing as "a prayer he wrote for himself," in the Epilogue she wrote for the Lowell Thomas biography of her husband. Sir Hubert had drawn from this Urantia Book passage: "To enjoy privilege without abuse, to have liberty without license, to possess power and steadfastly refuse to use it for self-aggrandizement—these are the marks of high civilization" (48:7.8).

Early Church Influence

Wilkins was brought up a strict Methodist, attending services every Sunday at the Mt. Bryan East Bible Christian Chapel, whose mission was to "spread Scriptural holiness over the land." He enjoyed singing in the church choir, and was allowed to learn the organ, but not the piano, which his devout parents regarded a sinful instrument. (His parents evidently allowed him to study other instruments as well, at the Elder Conservatorium of Music in Adelaide.)

Wilkins' sense of God as instilled in him by his parents and church was only intensified when he became an explorer. In a lecture delivered at Knox United Church in Edmonton, Canada, in 1938, during the time of his search for Levanevsky and his crew, he spoke of his consciousness of God:

> I think everyone will admit that those of us who go to the far corners of the earth cannot help but be God-conscious. We may not be God-fearing as such is preached in the churches. (Personally I don't think it is necessary to be God-fearing in order to be God-conscious.) It is not possible to travel all over this earth and see its magnificence and magnitude and not realize the existence of God. And when we travel in the lonely, desolate spaces of the polar regions we have time for contemplation. There we feel conscious of the greatness of God and more perhaps than we would if we spent our lives in the cities and in the rush of business. . . . Our minds, I believe, are more often turned to higher thought when relieved of all anxiety.

In the same lecture Wilkins told of the spiritual motive behind his desire to establish weather stations in the Arctic and Antarctica:

> I see on the distant horizon, as a result of the work we are doing in the polar regions, a vision of greater spiritual development. If it were not for that . . . I would not be spending my time in the frozen areas enduring what people consider to be hardships and privation.
>
> Sometimes, when people ask me why I go to the north and the south polar areas, I answer by saying that I want to know where the winds come from, where they are going, and what they are going to do when they get there. Wind is more than just the movement of the air. Wind carries moisture. It means life and growth, food and other things we

need. When we have simultaneously a knowledge of wind movements and its influence in various parts of the world, then I think mankind will find itself with greater opportunity for culture, leisure, and for consideration and development of what we call the higher things of life.

When we reach that point, I can visualize the possibility of greater leisure—and with it greater spiritual development.*

Beyond the Church

That Wilkins searched for knowledge along unorthodox lines is shown in his correspondence and in articles written about him.

In September 1926, while in Detroit preparing for the second Wilkins-Detroit Arctic Expedition, Wilkins received a letter from the Yogi Publication Society in answer to his request for a catalogue of Yogi literature. The Society recommended he begin his study of Yogi philosophy by reading the books listed in the order named: "Fourteen Lessons in Yogi Philosophy," "Advanced Course in Yogi Philosophy," "Raja Yoga," and "Chani Yoga."

In September 1933, in South Africa, while conducting a tour of the *Wyatt Earp* for two prominent Mormon leaders, Sir Hubert was presented with the Book of Mormon. An article that appeared in the September 29 issue of the local Mormon periodical described the meeting:[52]

> On Tuesday, September 26, 1933, I had the privilege of accompanying President Don Mack Dalton to visit Sir Hubert Wilkins. These are just a few notes that I took from their conversation . . .
>
> After talking a bit of Mormonism to Sir Wilkins, President Dalton gave him the Book of Mormon. He also explained to him that he should be very interested in the book especially as it tells how the Nephites and Jaredites built submarine barges and ships and crossed the oceans to the American continents in the years 2200 B.C., 600 B.C. and after.

---

*Wilkins' remarks here strongly resemble these Urantia Book passages:

"Civilization can never flourish, much less be established, until man has leisure to think, to plan, to imagine new and better ways of doing things" (81:2.14).

"Culture is never developed under conditions of poverty; leisure is essential to the progress of civilization. Individual character of moral and spiritual value may be acquired in the absence of material wealth, but a cultural civilization is only derived from those conditions of material prosperity which foster leisure combined with ambition" (81:6.6).

Sir Hubert is interested in submarines, winds and ocean currents. The explorer was extremely grateful to get "The Mormon Bible" as he called it. He said he had always wanted to read and study the Book of Mormon and he was so glad that he should be presented with one just when he would have the opportunity to read it. . . .

Upon our departure he shook our hands vigorously and again thanked President Dalton for the Book of Mormon and literature. As we were leaving the ship President Dalton said, "I hope that your trip will be a success and I am going to pray for you." His face took on new light and he said, "I pray and I believe in prayer because it helps me to guard against danger. I will be most grateful if you will remember me in your prayers."

In September 2, 1934, shortly before the Second Ellsworth Antarctic Mission, Wilkins wrote to Suzanne from Dunedin, New Zealand, telling her of a visit to a Spiritualist church:

I have just come from church—a Spiritualistic church. Just a little place in Dunedin and there were only about 30 people there. After the usual service a medium went to the platform and after telling two ladies about some personal affairs and giving them messages she gave me messages coincidentally from the same old man I think I told you I had described to me by a Spiritualist in London some five or six years ago—an old man with a beard. The medium said she could see a sparkling cross above my head and a blazing square about me. Many spirits were standing near all willing and eager to help.

The old man gave the same message—urging me to go ahead more firmly with my work and plans and assuring me of tremendous success, which the medium said would be sure to be mine. She also described one more or less like Eielson whom she said placed his hand on my shoulder and encouraged me to go on. She described him as wearing an Eskimo parka—although she did not know the name of the garment and she said that the man himself was a noble person. My mother she said was also there and said that while she knew I was not following the work she planned (she wanted me to be a miner), she realized that I could do more good if I went earnestly ahead and did what I was trying to do.

As usual the medium said she felt that I knew as well as she did and far better than most other people what was going on around me in this world as well as in the spiritual world and that from within my own self with the help of God I could accomplish all my desires. This was all given in an ordinary conversational tone in front of the whole congregation and a good deal more too most of which I could not plainly hear. . . . You should go to one of the Spiritualistic churches some time and see if a message comes thru for you.[53]

Harold Sherman

Harold Sherman played a pivotal role in Wilkins' extracurricular life of psychic research, involving him in publicized experiments in telepathy and introducing him to the as-yet-unpublished text of the Urantia Book.

Sir Hubert met Sherman sometime before 1931 when both were members of the City Club in New York. At the time Sherman was a best-selling author of boys' books and trying to make a name for himself as a playwright. The first evidence of a working partnership between Sherman and Wilkins is a musical-comedy radio play they wrote together in 1931 called "North Pole, Inc." Correspondence between them shows that Wilkins attempted to sell it to Hollywood producers, Sherman to New York radio networks.

Sherman also introduced him to the psychic circles in which he traveled. For instance, in a February 24, 1931, letter from Sherman in New York to a friend in Indiana, he wrote:

Just last Saturday evening I had Sir Hubert and Lady Wilkins out to the apartment, with a group of friends, to witness [psychic experimenter Jacques] Romano's mental feats. Romano relived Wilkins' life for him, avoiding the arctic adventures that are matters of world knowledge and tapping Sir Hubert's subconscious for personal incidents, names, dates and places that only Sir Hubert could know. . . .

On March 2, 1931, Sherman ended a letter to Wilkins:

It has been a pleasure to work with you and to become better acquainted with Lady Wilkins and yourself. It is a friendship which I hope I may always be able to deserve. We seem to have enough in common in our ideas on life to make future association hold promise of mutual benefit. . . .

Over the next few years, Wilkins and Sherman ran into each other from time to time, usually at the City Club. In the mid 1930s, while his wife Martha and daughters were living temporarily with relatives in Michigan, Sherman stayed at the Club attempting to sell his plays. In a letter to Martha dated August 17, 1936, Sherman wrote:

> . . . Dr. Strath-Gordon* came over at 3:30 P.M. and I had him meet Sir Hubert Wilkins who was just back from Europe. The fund of knowledge these two men exchanged between them in their discussion of geologic foundations the world over, the evidences of a lost continent of Atlantis, the climatic control of the earth's atmosphere, the proof of a race which built submarines and aerial vessels hundreds of years before the birth of Christ,** the possibility of another great earthquake which would make an inland sea of certain portions of America, the ability of both men to see the human aura in all its colors and to tell character and thoughts in this manner, the conditions leading to a world cataclysm which both men feel is surely coming—and the feeling that both men are to be used in the approaching world conflict and chaotic after-condition. . . .

Wilkins and Sherman continued to collaborate on their unsold "North Pole, Inc." play, spending most of the month of October 1936 rewriting and revising it, and attempting to sell it to the Columbia Broadcasting Network.

Thoughts Through Space

As mentioned in "Wilkins the Fact Finder," Sherman realized that Wilkins' upcoming mission to the Arctic, in 1937, to search for Levanevsky and his crew would provide an ideal opportunity for the two to conduct an experiment in thought transference. In the first chapter of *Thoughts Through Space,* Wilkins recalled his reaction to Sherman's proposal:

> The idea interested me. For many years, even in my boyhood, I had been conscious of the ability of certain individuals to receive directly

---

*City Club member and physician.

**Did Sir Hubert learn about the ancient submarine-building race from the copy of the Book of Mormon given to him in 1933?

> within their mind, without mechanical aid, impressions of conditions and events happening far away. . . .
>
> Through an acquaintance with Sir Arthur Conan Doyle, I had had some little experience with "spirit" demonstrations and clairvoyance, but I was not much interested in either of these phenomena. But I was interested in the possibility of dependable and willed thought transference from one living individual to another.
>
> If this quality could be acquired, or developed by individuals specializing in the subject, they, as specialists, might serve a useful purpose in the development of civilization. . . .

He added cryptically:

> I could give many records of astonishing predictions fitting exactly the happenings on my expeditions, but I am not particularly interested in those things which savor of clairvoyance or modern prophecy. These are phenomena which were not admitted by those from whom I received my early training—in spite of the fact that these very people were deeply religious, and admitted and accepted the genuineness of the prophecies detailed in the Holy Bible.
>
> I was interested in the possibility of "sensitives" who might respond directly to thought; for if such a quality were to be definitely and genuinely cultivated, that generally unused section of the brain might acquire an ability to serve humanity, and be instrumental in the development of human progress from barbarism toward refinement.

The apparent success of the Sherman-Wilkins telepathy experiments reinforced Wilkins' belief in the reality of the phenomenon.

The results were first published in the March 1939 issue of *Cosmopolitan* magazine, in an article by Inez Haynes Irwin titled "Some Call It Extra Sensory Perception." Sporadically throughout 1939 and 1940 Sherman and Wilkins continued to work on the manuscript which would result in their joint book, *Thoughts Through Space.*

On December 12, 1940, they reenacted their experiment on an hour-long radio broadcast, "Strange as It Seems," which resulted in a record-breaking volume of mail. By the middle of March 1941 they were ready to submit their manuscript to publishers, and in 1942 *Thoughts Through Space* was published.

Urantia

[UCA]

Sir Hubert Wilkins learned of the Urantia revelation sometime in 1941 from Harold Sherman, who had heard about it from a former Chicago detective named Harry Loose.

Sherman first met Loose in July 1921 in Marion, Indiana, after Loose gave a talk on "Crime and Criminology" as a Redpath-Chautauqua lecturer. Sherman, then a young reporter for the *Marion Chronicle*, interviewed him. The two spent an evening in Loose's hotel room, during which Loose purportedly communicated telepathically with his wife and exhibited other extraordinary powers of mind. Loose assured Sherman that he too was psychically gifted and would have a tremendously important role to play in the world. He further told Sherman that he, Loose, would be disappearing for a while, "while on a mission."

In February 1941, twenty years after their first and only meeting, Sherman tracked Loose down, then retired and living in Southern California, and wrote him a letter. Loose responded immediately. Among other things, he told Sherman:

> Remember to watch for a tremendous book which will be published in about two years. It has now been thirty-five years in the building. It is not mine but I had something to do with it. You will recognize it when it appears. It will be world shocking. It will clarify so very much that is cloudy in our present-day Bible. It is a true spiritual revelation to this age written by Intelligences that have never been earthbound and who have to do with the governing of this tiny earth in this very limited universe. Please believe every astonishing word. It is the *Truth*. I *know*.

Loose and his wife Emily had been members of the Forum, a private group set up to study the Urantia material at the home of its "custodian," Chicago physician William S. Sadler. Sadler claimed to be in direct contact with the superhuman revelators, who communicated through a man he declined to identify. Also privy to these transactions were Sadler's wife, Dr. Lena Kellogg Sadler; their son, Bill; Lena's sister Anna and her husband, Wilfred Kellogg; and family friend Emma "Christy" Christensen. The group called itself "the Contact Commission."

The Looses left Chicago in about 1934, but stayed in touch with members of the Forum who apparently kept them abreast of the evolving Urantia manuscript.

The Urantia Forum at their annual picnic in 1934 [KATHARINE HARRIES]

A voluminous correspondence developed between Loose in Monterey Park, California, and Sherman in New York. Loose became Sherman's psychic mentor and coach, sharing content from the Urantia text mixed in with his own metaphysical beliefs and claims. Sherman and his wife, Martha, were thrilled with what they were learning.

Sherman no doubt mentioned Loose to Sir Hubert Wilkins, with whom he was in regular contact, as they were working together on the manuscript of *Thoughts Through Space*. In a March 19, 1941, letter to Loose, discussing who could join him in the world-betterment mission which Harry Loose was helping him conceive, Sherman wrote:

The Contact Commission

Above: Christy, Lena Sadler, Dr. William S. Sadler, Bill Sadler, circa early 1930s [UBF]
Right: Dr. Sadler in younger days [UBF]
Below: Anna and Wilfred Kellogg at an annual Urantia Forum picnic, circa 1943 [UBF]

> Sir Hubert Wilkins is another highly developed soul with whom I could go a long way in matters of this kind. He has said to me that he feels the universe is filled with Intelligence of infinitely high and low grade, in a state of unfoldment.

Loose encouraged the Shermans to meet Dr. Sadler and join the Forum, which they did in July 1941. Sherman had earlier been hired by Warner Bros. to write the screenplay for *The Adventures of Mark Twain*; this gave the Shermans the opportunity to live in Hollywood near Harry Loose from August 1941 to May 1942.

Joining the Urantia Forum

On October 27, 1941, Sherman wrote to Dr. Sadler:

> I wish that Sir Hubert Wilkins might be added to your Group. I am sure he is one of us and his name would add greatly because of his interest and championing of these higher mental powers. . . . Might I have permission to ask him to call upon you when he is next in Chicago? He is one of the finest souls it has ever been my privilege to meet.

In early March 1942, on a joint book tour for the newly published *Thoughts Through Space*, Sherman introduced Wilkins to Sadler and the Forum, in Sadler's home at 533 Diversey Parkway. Both Harold and Sir Hubert read papers (i.e. chapters from the Urantia manuscript) as time permitted. On March 8 they attended their first Forum meeting and each gave a short talk to the group. Later that day they had Sunday dinner with Dr. Sadler and Christy. (Lena Sadler had died in 1939.)

Excerpts from the Wilkins-Sherman correspondence show Wilkins' enthusiasm for the Urantia manuscript and the Forum. On March 10, 1942, he wrote to Sherman:

> It was good to see you again and I will be forever grateful to you for the introduction to the Forum. It seems that I will have an opportunity between April 21 and sometime in May to visit Chicago for several days and I will do so in order to gain more knowledge of the Great Universe. I am writing to Mr. Loose with this mail. Also to Dr. Sadler.

Dr. Sadler's residence in Chicago where the Papers could be read [UCA]

Sadler was similarly impressed by Wilkins. In his March 17, 1942 letter to Sherman, Sadler wrote:

> We were all glad that you and Sir Hubert could be with us, and I assure you that everyone is sincerely charmed with Wilkins. It seems to me he has about all of those things which we like in the British, while he has managed to escape those that we Americans are not so fond of.
>
> I am writing him today in answer to the sweet little note he sent about his visit to Chicago, assuring him of our pleasure in seeing him at any and all times it is possible for him to make Chicago.

Sherman to Wilkins, March 17, 1942:

> I still get a glow from the memory of our weekend in Chicago. And I am so happy that you feel you have found an "anchor" in the Group in Chicago. . . . You made a splendid impression upon the Group and the Sadlers. Bill Junior told me on Monday night, after I had read most of the day (and what reading!), that he had wept after our departure Sunday night, from joy at our having come to them—looking back as he had upon the long, painstaking seventeen years of preparation; and now to have the realization that the time was at last approaching

(perhaps still a year or two away) when this knowledge would be made known to the world, and might have people like yourself in back of it . . . . I am glad you can plan to be in Chicago for a few days in April.

Wilkins to Sherman, April 18, 1942:

I have been reading steadily from 9:30 to 12:30 and 1:30 to 5 p.m. every day Monday to Saturday. It is grand to be able to keep at it and "hold" the picture in mind—although I've had some worry in the writing, in the evenings, of a new lecture, "Air Supremacy and the Control of the Pacific" which I must give at Detroit Economic Club on Monday.

Wilkins to Harry Loose, April 18, 1942:

Dear Mr. Loose:

This has been a most enlightening and inspiring week for me. I have spent from Monday until today reading the papers and have "skimmed through" those up to No. 97.

What a grand and glorious picture the first section presents, and in the geological, anthropological and theological sections I have found

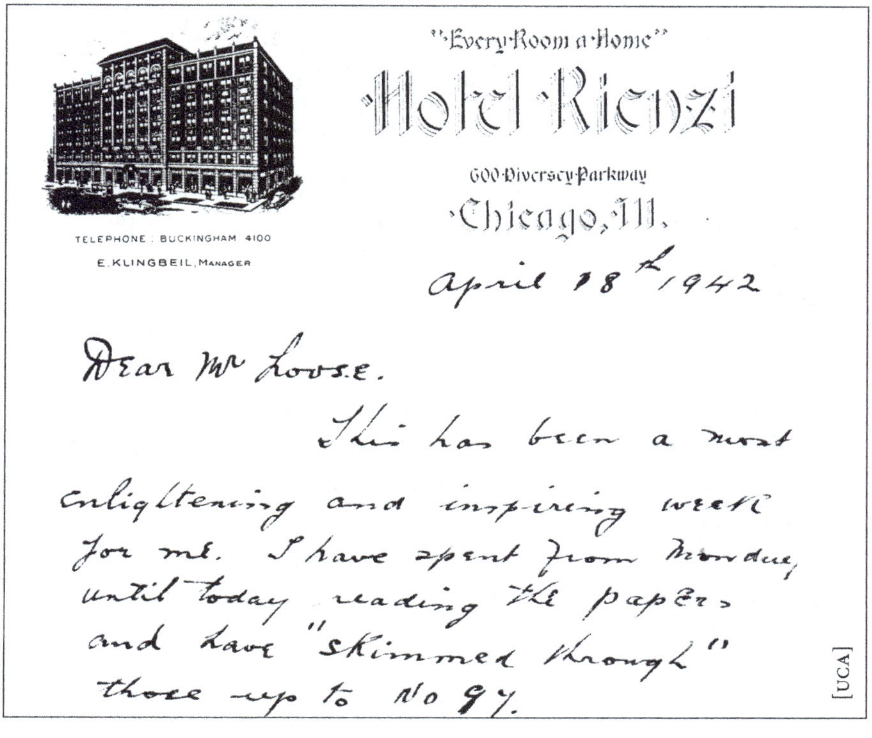

"Every Room a Home"
Hotel Rienzi
600 Diversey Parkway
Chicago, Ill.

TELEPHONE: BUCKINGHAM 4100
E. KLINGBEIL, MANAGER

April 18th 1942

Dear Mr Loose.
This has been a most enlightening and inspiring week for me. I have spent from Monday until today reading the papers and have "skimmed through" those up to No 97.

[UCA]

not only those matters I had previously met but a host of other matters which makes the picture so much more complete and clear. What astounding possibilities are presented! I have to leave here tonight for Detroit and must go to Green Bay, Wisconsin, on Tuesday but I expect to be back at this hotel on Wednesday next and on Thursday take up the reading again. I am looking forward with great eagerness to the reading of the other papers and hope, after I have read all of them once, to have time to more leisurely go through the parts which will most constructively prepare me for service and tend to *my own development*—a matter very obviously needed.

I have spent only one evening this week with the Dr and his son and I have not mentioned you and *will not* unless you indicate that such is desirable. I am, however, looking forward to the time when I may, if possible, have the opportunity of learning some of the truths from you personally. With best regards to you and your family,

Sincerely,
Hubert Wilkins

In mid May 1942, Sir Hubert wrote a letter to Sherman, who shared it with Wilfred Kellogg. Kellogg asked permission to copy some of it as he felt that Wilkins' estimate of the Urantia papers was "so concise and clearly put," he wanted it on record. Wilkins had written:

Until the last minute before rushing to catch the train to New York, I was reading and just managed to get through the last chapter. The whole thing is so immense and astounding that one is lost for words to describe it. It is, however, likely to be subject to criticism by the established churches, both in respect to its origin and to its revelations which deprive some of the sects of the basis of their foundation. To me, however, it is the first connected, clear exposition of *the* "faith"—the need for faith and the reward of faith, a combination of considerations which seem to confront Man no matter what race or color, and notwithstanding his stature of "education."

We might be intrigued in the matter of faith because of religious training very early in our youth. But the wish to *know* something of the cosmic organization seems paramount in all humans, "heathen" or otherwise.

> The papers will therefore fill a long-felt want.
>
> They will need—as have other writings—a good deal of *expounding* if they are to reach the multitudes, but that has been found necessary in respect to all "teachings." Yet the Papers in themselves are more *absorbable*—if one has much capacity—as a whole, than any other theological tome.
>
> I will arrive at Chicago on the 21st and spend two or three weeks in reading again the material. After that time I will probably get back to the farm (in Pennsylvania) until about August when I will probably go to Alaska . . .

The Hollywood assignment finished, the Sherman family moved to Chicago in early May to make a study of the Urantia teachings. On the evening of May 21, 1942, Harold and Martha were joined by Sir Hubert, and the three were guests of Dr. Sadler and Christy for dinner followed by an evening of conversation. They attended their second Forum meeting on May 24, and Sherman wrote to Loose the next day:

> Wilkins is delighted beyond words and reiterates that he would always be eternally grateful to us for this association. We are reading in the front second-floor room, Martha and I at one end and Wilkins at the other. We read out loud but it doesn't disturb him since he is deaf.*

On June 6 the Shermans and Wilkins attended the annual Forum picnic, being seated at the Doctor's table in places of honor. On June 8, Sherman wrote to Loose:

> The entire membership is in love with Wilkins, who has made a tremendous impression. He had to leave after the picnic and we drove him to the station to catch a midnight train for Washington, where he has been called on the Alaskan situation. Wilkins told me he never could repay *us* in this lifetime for putting him in touch with the Forum and this great knowledge which he has no doubt comes from "higher intelligences," the most amazing document ever to come to earth. You should be hearing from him soon. How he has dug into this material—long hours every day for weeks!

---

*Wilkins wore a hearing aid.

Wilkins, back in Washington, D.C., to Sherman, June 14, 1942:

I miss the daily readings and the informative and enjoyable conversations with you and your charming wife.

Sherman to Wilkins, June 16, 1942:

You are much missed here—by everyone—and especially by Martha and me.

We have now reached the Jesus papers, and have had many "moving experiences." We are commencing to realize, more than ever before, that we are among the most *privileged* of humans—as you have so expressed—and this must imply a great spiritual responsibility in the time to come.

What a *perspective* is now added to our mental and spiritual horizons! Never again can we think or view things as before. It is all, as you have also said, absolutely indescribable.

Wilkins to Sherman, June 30, 1942:

Still trying to recall to memory daily what I can of the "material" but as you indicate it is a colossal job.

And on July 15, 1942:

Congratulations on your 44th. Since reading the papers and reflecting on my 54th this coming October, I have an entirely different attitude toward Earth time and human life and birthdays.

Some of Wilkins' Observations

After their honeymoon period in the Forum, Harold and Martha Sherman began to question Dr. Sadler's and his son's plans to privately publish the Urantia Book and to form an official quasi-religious organization around the book after publication. In the summer and fall of 1942, Harold shared his concerns with the Forumites, and those who agreed with him wrote a petition calling for open discussion of these matters. Forty-eight members signed the petition and a group of three delivered it to Dr. Sadler on September 4. This action infuriated the doctor and caused a temporary division among Forumites. Sir Hubert had signed the petition but tried to remain neutral in the ensuing controversy.

Harold and Martha Sherman in Chicago, 1943 [SHERMAN FAMILY]

On Harry Loose's suggestion, the Shermans began to keep diaries recording Forum events.* Wilkins' occasional visits to Chicago and his observations on Urantia matters are recorded there. On November 10, 1942, the Shermans wrote:

> We all reiterated our faith in the contents of the Book of Urantia, and Wilkins declared that he felt no human explanation of its origin was necessary—simply the statement that "here it is," and leaving it to the

*These diaries, spanning the years 1942-1947, have been published as *The Sherman Diaries* by Square Circles Publishing (since revised and republished as *The Urantia Diaries of Harold and Martha Sherman*).

judgment and appraisal of individual readers to evaluate its spiritual and revelatory authenticity. In this connection he further declared that he had never liked this "spiritual mystery and secrecy" anyway, and did not think it was in keeping with the intent of higher spiritual beings and the vital interest they must have in the destiny of human creatures.

May 7, 1944:

Sir Hubert Wilkins was here today and had lunch with us. . . . Wilkins said he had reread the "neighboring planet" paper and also the papers on the mansion worlds and found them as absorbing as ever. He had read from 11:00 to 5:30 yesterday and returned to Washington tonight.

April 6, 1945:

Sir Hubert surprised us with a phone call this Friday a.m. stating he had recently returned from the Aleutians and had come into Chicago from Dayton to stay until Sunday noon so he could do some reading. We had dinner with him Friday evening and he told us he was reading Book No. 1 including the Foreword, all of which was now set up in page proofs. Sir Hubert again stated that this material was obviously beyond human ability to create or even to actually conceive. He wondered how average humans could ever begin to comprehend it or even to become interested. He said few Forumites were qualified to understand it, much less to teach it to others.

October 25, 1945:

Hubert Wilkins arrived unexpectedly en route to the Coast and had lunch with Harold. He had spent the morning reading the papers about the nature of the atom and it was his opinion that the papers still contained advance information on the elemental portions of the universe. Wilkins said he had never found anything contrary to scientific fact in the papers and regretted, again, his inability to reside in Chicago and devote his full time to a study of the material. He said he was planning to return within three weeks and spend some weeks here.

November 11, 1945:

We were surprised to find Sir Hubert standing on the O'Connell's [restaurant] corner waiting for us when we crossed over to 533 today. He had arrived by train from New Orleans and apparently aimed at attending the Sunday meeting. He is to be here for two weeks.

Bill read the second half of the Foreword, stopping to ask every few minutes with his tongue in his cheek if we had any questions. The profundity of this paper is undoubted and Sir Hubert said it could never be comprehended by any human, not even Einstein, but that its very profundity might help give the Papers an added authenticity. As for our own comprehension, Sir Hubert said we were still billions of years away.

The Shermans' diary entry for November 14, 1945, mentions the Urantia notebook Sir Hubert had just started keeping:

We had dinner this night with Sir Hubert who told of an evening's discussion with Bill who hoped to teach the Urantia philosophy when the Book was out. . . .

Wilkins explained he was making notes for the first time in a rather original way. He is listing under such headings as Man, Mind, Spirit, Thought Adjuster, and other headings, what he considers to be the outstanding facts, so that he can crystallize this knowledge in his own mind. He said he has found nothing as yet which is contradictory in this whole voluminous material.

November 18, 1945:

We had a pleasant evening with Sir Hubert, Mrs. Gusler, Elsie and Almond,* discussing the Papers. . . . Wilkins had somehow thought, all this time, that the Papers had been dictated by direct voice through the subject and taken down by Christy. He was put right on this point by both Dr. Sadler and Bill, who declared that all papers were written in the handwriting of the sleeping subject. Wilkins asked to see the original papers and particularly those which had been altered although locked in the safe (according to the Doctor's story), but Dr. Sadler said no one outside the contact commissioners would be permitted to see these papers, and they would be burned or dematerialized when the Book was ready for publication.

Almond asked why reputable scientific witnesses hadn't been called in long ago to help substantiate these revelations, and Wilkins said he didn't know but that the subject matter would speak for itself. Almond

*Forumites Rachel Gusler, Elsie Baumgartner, and Almond Fairfield, with whom the Shermans had become close.

then said, near the close of the evening, "So far as I am concerned it is all a beautiful fairy tale—nice if true." This comment did not seem to disturb Sir Hubert, who is doing exhaustive research and is deeply engrossed in the material.

November 23, 1945:

We discussed further Urantia matters at dinner with Sir Hubert while his mind was fresh on the papers he had read. He said a description of the mansion worlds and their exact location was given in Paper 47, which was news to us. Bill had formerly stated that he didn't know where the mansion worlds were situated. Wilkins asserts that the Papers say the seven mansion worlds are the out-satellite groups around the first satellite surrounding Jerusem. This first satellite is the reception world and is ten times the size of the morontia mansion worlds, which in turn are each about the size of our earth.

To our surprise, Wilkins stated that Lucifer and his associates were confined on a prison world, the sub-satellite connected with the seventh satellite (of seven) surrounding Jerusem. We had been led to believe that the moon was being used as this prison world. . . .

Wilkins remarked that the latter papers on the Thought Adjuster were not written with the same majesty and authority as the former, and that it seemed they could easily have been written by some human although indited by a Mighty Messenger. Wilkins speculated that it might have been considered necessary to simplify and harmonize some of the Thought Adjuster papers for mass consumption. . . .

Wilkins reported that he was making this outline of the Papers, not in his words but by direct quotes of paragraphs, since he did not want to risk misinterpretation.

December 1, 1945:

Wilkins was here for dinner and the evening. . . . Wilkins remarked that there were very few people he felt he could recommend for membership, the more he got into the Papers.

In discussing psychic matters, Wilkins stated that the Papers declared emphatically there was no return of the entity to this planet after death until that entity had progressed through the Havona circuits or after the end of a dispensation. . . .

Sir Hubert's Concept of the Universal Plan

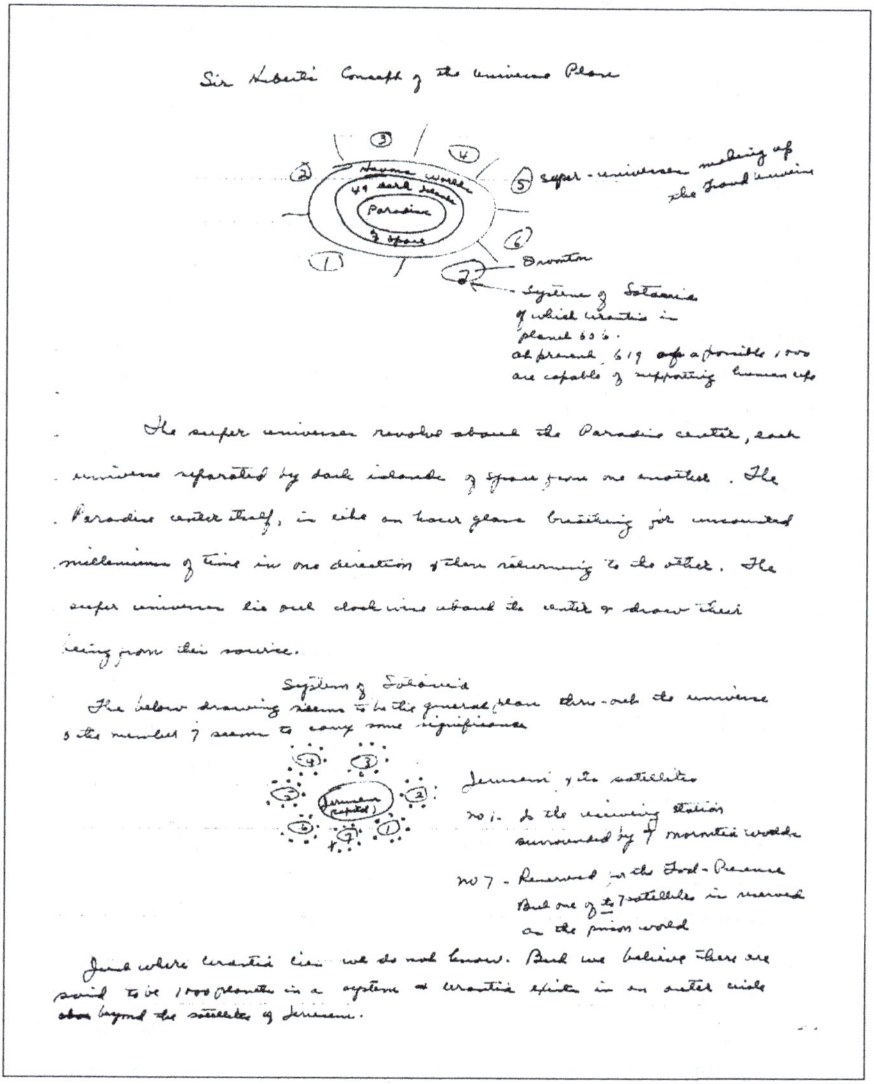
Sir Hubert's Concept of the Universe Plan

Havona worlds
41 dark islands
Paradise
7 space
super-universes making up the Grand universe
Orvonton
System of Satania of which Urantia is planet 606.
At present 619 out of a possible 1000 are capable of supporting human life

The super universes revolve about the Paradise center, each universe separated by dark islands of space from one another. The Paradise center itself, is like an hour glass breathing for uncounted millenniums of time in one direction & then returning to the other. The super universes lie out clock wise about the center & draw their being from this source.

System of Satania
The below drawing seems to be the general plan through-out the universe & the number 7 seems to carry some significance

Jerusem (capital)

Jerusem & its satellites
No 1 - Is the receiving station surrounded by 7 morontia worlds
No 7 - Reserved for the Lord-Presence
But one of its 7 satellites is reserved as the prison world

Just where Urantia lies we do not know. But we believe there are said to be 1000 planets in a system & Urantia exists in an outer circle beyond the satellites of Jerusem.

A page from the diaries of Harold and Martha Sherman, in Martha's handwriting (The Sherman Diaries,* Volume 4, *p. 395)* [UCA]

*All six volumes have since been revised and republished as *The Urantia Diaries of Harold and Martha Sherman* (Square Circles Publishing).

It occurred to Martha that Jesus had returned after death, and why should he have set such an example were return not possible? . . . Wilkins could not answer this point and apparently is keeping an open mind, for he is quite willing to go on a proposed radio program* and express his opinions about possible spirit return and other phenomena. . . .

Wilkins significantly remarked that Jesus' ministry only lasted about a year and a half and that a human could perform a service and become world-renowned, if necessary, within a brief period, after years of quiet preparation. Wilkins reiterated that he had been eager to go on to the next life since a boy and there had been many times in his experience when it would have been easier to die than to live, but he had persisted because he felt he had a mission of some kind to perform. Some tremendous urge is certainly working in him now.

March 14, 1946:

Hubert Wilkins suddenly appeared on the scene and spent the evening talking to Harold . . . . Harold asked Sir Hubert what he thought of the style and content of the Jesus papers, and Sir Hubert said that he did not feel they maintained the same quality as the other Urantia papers. He said this disturbed him at first, and he spoke to Dr. Sadler about it, who explained that the Jesus story was largely for the consumption of children, who would "love it." Dr. Sadler said that was why the story was written so simply. . . .

Hubert said he had finally decided that Jesus' one important message for humanity was, "In my Father's house are many mansions," and the promise that there was continuity of life beyond death and training worlds for soul development. To Sir Hubert, this seemed to explain why Jesus refused every offer to be set up as a teacher, to be exalted or to head any religious organization. Jesus wanted to avoid being idolized or worshipped, and wished humans of all ages to value his message and not the man.

Hubert admitted that Jesus' boyhood lacked conviction when the statement was made that "no young man who ever lived had such trials and decisions to make or such obstacles to overcome as Jesus of Nazareth." Hubert said that millions of young men had been forced to assume much greater responsibilities, under more severe hardships, in

*Their "Mysteries of the Mind" radio program, which never materialized.

> worse environments than Jesus, and millions of men have suffered and died even more horrible deaths. The Jesus papers, then, do not make a good or convincing case in this regard.
>
> Harold called Wilkins' attention to inconsistencies as previously listed here and also mentioned that these [Jesus] papers were unsigned as to authorship. Hubert did not seem much disturbed about this and said he did not think these papers represented the work of the Supreme Intelligence or even too-high intelligences beyond the human, and that they could not be taken, in all instances, as the final word on every subject. Apparently Hubert is inclined, for the most part, to be uncritical and not raise any issues but to extract what he feels is of value from the Papers for his own use and development.

In May 1947, the Shermans, estranged from the Sadlers and most of the Forum members and convinced that Sadler had "doctored" the text with his own beliefs and prejudices, left Chicago and moved to Arkansas to pursue other interests. Wilkins, however, remained committed to the Urantia revelation and continued to visit the Forum in Chicago as often as he could, keeping the Shermans informed of Urantia developments via their correspondence.

On October 25, 1950, the newly formed Urantia Foundation sent a letter to Forum members asking for contributions towards the approximately $45,000 needed to print ten thousand copies of the Urantia Book. Sir Hubert responded with a check for $1,000.

Dr. Sadler conveyed his appreciation in a letter to Wilkins on December 15, 1950:

> We were all greatly relieved to get your letter from New York because a couple of weeks ago the rumor was among us that you had met with an accident, and not knowing just where to write you we had been waiting to learn more or to hear from you and thus find out the trouble. . . .
>
> Of course your letter had another and additional bit of charm in that it advised us of your contribution of $1,000 to the Urantia Foundation. You can be sure we greatly appreciate your interest and cooperation. If I am informed up to date, I think your contribution is the one that brings the total amount up to $40,000 with a little more than half the membership of the Forum heard from.

The Published Book

[SP]

In a March 16, 1955, letter to Sherman, Wilkins wrote:

> Now, real news, which you may already have heard. The Urantia Book will be off the press by September 1, 1955—10,000 copies. From what I hear there will be no effort to publicize it; on the contrary, as a matter of fact. As I understand it, members will be authorized to purchase a few copies for themselves and their friends with the idea of forming various and sundry sections of the Urantia Brotherhood. . . . Dr. Sadler, when I saw him on March 8, was again looking well. Bill was very busy—home for a few hours between trips. And Mr. Kellogg is getting rather weak and shaky—working too hard on the Index, I believe.

The Urantia Book was finally published on October 12, 1955. It consisted of 2,097 pages comprising 196 papers, or chapters, describing the nature of God, the structure of the universe and its divinely planned development, other worlds, angels, life after death, the origin and history of planet Earth—known as Urantia—the life and teachings of Jesus, and Man's eternal destiny as a perfected citizen of a friendly universe. It gave a picture of reality in which science, religion, and philosophy were unified. The authors, as claimed within the book, were a corps of celestial beings authorized to reveal this information to humans, channeled through a favorably constituted human conduit known as the "contact personality" or "sleeping subject."

No sooner was the book off the press than Wilkins began giving away copies to friends and colleagues, attaching a cover letter with each book. One such letter accompanied the volume he gave to Winston Ross, the Wilkinses' long-time personal secretary, on November 1, 1955:

> Dear Winston,
>
> . . . For many years I have been associated with a group in Chicago which has been interested in publishing some papers of material revealed to us by visitors from the outer universes. At last we have been able to print and privately distribute the Book and I would like for you to have a copy.
>
> At present we are not telling many of the recipients of the manner in which the information was received but I can tell you for your own information that the texts of the papers were spoken by the revelators through a man in his sleep and who to this day has no idea he was the medium.* Learning that this man was "talking in his sleep," it was arranged to have a stenographer record the statements and soon it was possible for those concerned not only to listen and record but also talk with the revelators as you and I might talk.
>
> The mass of information of the Book is at first bewildering. To most of us it came piece by piece and was not so overwhelming.
>
> How much of it you and your mother might accept of the information I do not know, but if you can accept it, the information as to the possibilities of survival after death and the experiences thereafter, as mentioned in the paper "Morontia Life" and elsewhere, is most inspiring and comforting.
>
> The Book is so heavy that it might be necessary for you to get a bed-reading stand so that your mother, if she is interested, can have it supported before her as she reclines in bed. I have made such a stand for myself and find it very helpful. . . .

*Wilkins apparently went back to thinking that the Papers had been orally delivered, not through the handwriting of the man.

To Sherman he wrote on November 4:

> At last the Book is being distributed and I have received my copy.... [It] is, as it has always been, bewildering unless one takes it leisurely. There is a lesson in almost every sentence but some of it is probably beyond the scope of many people. I would have preferred to have it in several separately bound volumes—it is too heavy to cart around, and you need a table to put it on for reading. It is, however, a handsome publication and the first printing will, I believe, some day be museum pieces....

The Black Notebook

Forumites had been encouraged to keep notes, their notebooks not to be removed from the premises at 533 Diversey Parkway until after book publication. As mentioned above, Wilkins had begun to make notes in 1945, copying down passages that interested him, grouping them alphabetically according to topics.

On October 27, 1955, Wilkins wrote to Bill Sadler:

> And now that the material has been released in printed form I would like to have someone post to me the Notebook I used for reading during my occasional visits. The last time I was in the Reading Room the Notebook was in the lowest drawer, on the right-hand side of the right-hand bookcase as you faced the bookcases. It is a black-covered loose-leafed book and has my name on it. In it various subjects are segregated.

On November 3, Bill Sadler replied:

> As this letter is being written, Leone* has dug out your old Notebook and it will be posted to you this week.

On November 28, after Sir Hubert notified the 533 office that he hadn't received the Notebook, Leone responded:

> I, personally, mailed your Notebook to you at your New York address (37 W. 53rd St.) about ten days or two weeks ago. Because of its nature it was necessary to mail it first class, but not necessary to insure it, so I have no receipt to give me the exact date.... Please accept my apology for its delay. I am sure you will receive it but will you please let me know

---

*Leone Gill Sadler, Bill Sadler's first wife, who was helping in the office.

> when you do. It looked like such a wonderful Notebook and I would despair of having to reproduce it.

Eventually it did arrive but it would be many decades before it resurfaced. (See p. 105 for the story.)

Personal Evangelism

Sir Hubert spent much of the last few years of his life studying the Urantia Book and trying to interest his friends and colleagues in it. Realizing that the book was too demanding for the average reader, he was undaunted by the lack of positive reaction. In a letter to Harold Sherman from December 30, 1955, Wilkins wrote:

> Of the 12 Urantia Books I have "distributed" only 1 interested response. Some seemed to think it is a joke—a novel—or something of the kind. Anyway the response is a good criterion of their real mindal ability.

On April 8, 1957 he wrote to Sherman, reporting more interest:

> The URANTIA BOOK is also in great demand by my friends. I have disposed of some 25 and have a request list waiting to be filled. Some recipients are thrilled, some indifferent, and what seems strange, but not too strange—from what we know, some of the books just seem to have disappeared in the Post Office and tracers have found no evidence of them.

On his last visit to Antarctica in 1958, Wilkins took along his Urantia Book, marking what was certainly the first and possibly the only time the book has reached a polar region. Helicopter pilot James Waldron, who shared a room in Little America with Sir Hubert for two weeks during the trip, commented:

> He was a very old man, too old it seemed to be on the Ice. Since he died a year or so after this visit to the Continent it only proved further that he shouldn't have been there. I got to talk several times with Sir Hubert and he proved to be a very interesting person, however, a very strange man. . . . He always seemed to be reading from a Bible sized book and when I questioned him about it he told me that he belonged to a "religion" that welcomed only a handful of members from each world throughout the

> universe [*sic*]. He considered himself to be one of the fortunate few who had been selected to belong to this extraordinary cult. He believed that the various members of this "religion" communicated telepathically over the vastness of space [*sic*]. He was more than convinced that this "religion" was no hoax. Of course, I didn't challenge his beliefs.

Remembering Sir Hubert

Sir Hubert died on December 1, 1958, shortly after his trip to the Antarctic, and articles about him soon appeared in newspapers and magazines. One such piece, "The Most Unforgettable Character I've Met," was published in the July 1959 issue of *The Reader's Digest*. It was written by Vilhjalmur Stefansson, the explorer who led the Canadian Arctic Expedition in 1913, in which Wilkins served as photographer. Stefansson told of Wilkins' many expeditions and praised him for his "astonishing vigor" and the "unique warmth that [he] radiated. He possessed a mystical magnetism and was always acutely sensitive to others' needs and wishes."

One person who read the article was Dr. William S. Sadler. Sadler had noticed the lack of mention of Wilkins' involvement with the Urantia revelation and on August 14, 1959, he wrote to Nancy Hildick, Stefansson's secretary at Dartmouth College (home of The Stefansson Collection on Polar Exploration), to provide information on this aspect of Wilkins' life:

> What I wish to bring to your attention is the matter of Sir Hubert's religious inclinations for the past twenty-five years [*sic*], and as a background for this, I am sending you a copy of The Urantia Book, published three or four years ago.
>
> About twenty-five years ago Sir Hubert learned about a group of us here in Chicago who were attendant upon a rather unusual psychic phenomenon, and he became a member of this group. During a period of about twenty-five years we received communications through this subject, which purported to be superhuman in origin, and these communications were subsequently published in The Urantia Book, which I am sending to you.
>
> For the last twenty-five years this was Sir Hubert's personal religious philosophy. He came to see us every year or two, spending two or three

weeks reading, studying, and making notes on these Papers up to the time of the publishing of the Book in October, 1955.

During these annual or biennial visits he spent about ten hours a day studying the Urantia Papers. During these visits it was his custom to spend several evenings at my home during which we engaged in very deliberate discussions of the teachings of the Urantia Papers. Sir Hubert Wilkins was a member of this group, which we, in those days, called THE FORUM, but which now functions as THE FIRST URANTIA SOCIETY.

In 1954 [*sic*] when permission was granted for publication of the Book a letter was sent to about 250 members of this original Chicago group, asking for contributions of $50,000.00 for the publication of 10,000 copies. Sir Hubert's response to this letter was the first money received [*sic*] in the office of the Urantia Foundation, and it consisted of a contribution of $1,000.00. . . .

I happen to personally know that Sir Hubert was profoundly interested in the text of this Book and that it represented his personal beliefs for the past twenty-five years.

Before the publication of the Book, no written matter was permitted to be carried away from the address which represented the custodianship of the Papers. The notes which Sir Hubert made, during his annual or biennial periods of study, were sent to him upon publication of the Book.

I do not know how much Mrs. Wilkins knew about her husband's connections with this group, but when I read Dr. Stefansson's article in *The Reader's Digest* I thought this was a matter of too much importance to be left out of the record of Sir Hubert's life. Knowing how much he valued the Urantia Papers, I thought they ought to be part of the archives which would communicate the extraordinary life activities of this unique man.

Sadler's account of Sir Hubert's Urantia involvement contains a number of inaccuracies and omissions. He says no fewer than five times that Wilkins had been reading the Urantia text for twenty-five years, when in fact he had been introduced to it in the spring of 1942, just sixteen-and-a-half years before his death. Sadler also misremembered a minor detail about Wilkins' $1,000 contribution to the Urantia Book publication fund; as recorded above, the fund drive began in 1950 (not 1954) and Wilkins' contribution was not the first to be received, but among the last. More importantly, Sadler omits mention of Harold Sherman, who brought Wilkins into the Forum and who shared Wilkins' abiding interest and belief in extrasensory perception.

In 1960, Sadler was contacted by Vincent Buranelli, a representative of Lowell Thomas, the writer and radio broadcaster who had flown with Wilkins and others across the Arctic in 1957. Thomas was preparing the first American biography of Sir Hubert and had Buranelli send Sadler the draft of a chapter entitled "Adventures of the Mind." It dealt mainly with Wilkins' and Sherman's ESP experiments but also mentioned Wilkins' attachment to the Urantia Book. After reading the draft, Sadler replied, in his June 30, 1960 letter to Buranelli:

1. On Page 490, I would suggest the omission of my name. There are many reasons for this, not the least of which is that men in my profession are supposed to avoid personal publicity. The text would therefore read, beginning about the third line from the top, "and long conversations with its members and the leaders of the movement."
2. We dislike very much to see the Urantia movement mentioned in connection with telepathy and extrasensory perception. These psychological matters are utterly foreign to the Urantia Book and the Urantia movement.

Sadler continued:

> I feel certain that Sir Hubert did not himself put the Urantia Book in context with these psychological phenomena. This chapter, it would seem to me, had been arranged from data which he had prepared, but had not arranged in this order. We have no criticism to make of the text as it appeared in this chapter, as regards the Urantia movement, except in this matter of unfortunate contextual arrangement.
>
> We would be pleased and most happy if you could see your way clear to take all reference to the Urantia Book and the Urantia movement out of this chapter dealing wholly with telepathy and find for it a place in some other less objectionable context.

Sadler apparently assumed, or wanted to believe, that Sir Hubert had given up his interest in psychic phenomena after embracing the Urantia revelation. But the Wilkins-Sherman correspondence shows that the two were still trying to interest producers in proposed radio programs dealing with psychic phenomena well into the 1940s and 1950s. That Wilkins did not share Sadler's aversion to extrasensory perception, and that he indeed stood by the authenticity of his and Sherman's 1937 telepathy tests, is shown in his introduction to the 1951 reissue of *Thoughts Through Space.* Alluding to information in the Urantia Book regarding Adam and Eve's use of telepathy, Wilkins wrote:

> There has been a far greater span of time since the first demonstration of Extra-Sensory Perception, or thought transference, and the recordings of Harold Sherman than the interval between 1913 and 1951. Adam may have used thought transference to communicate with Eve and it seems likely that the wisdom of the Serpent was not entirely conveyed to Eve by means of human colloquial dialogue.

The "Adventures of the Mind" chapter was scrapped from the Lowell Thomas biography. Whether Sadler's objections played a role in the decision not to include it remains a mystery.

Another biographer, John Grierson, did touch upon Sir Hubert's Urantia and ESP involvements, in his book, *Sir Hubert Wilkins: Enigma of Exploration,* published in Great Britain in 1960. He mentioned *Thoughts Through Space,* "published jointly by Sir Hubert and H.J. Shurman [*sic*]," and erroneously described the Urantia Book as "the religious work of many authors of whom Wilkins was one. 'Urantia' means Our Universe [*sic*]." Grierson apparently learned of the Urantia Book and Sir Hubert's attachment to it through Suzanne Wilkins or through the Wilkinses' secretary, Winston Ross, because he quoted from Wilkins' November 1, 1955, letter to Ross, reproduced above.

Grierson added:

> What should one make of Wilkins' avowed connection with the Urantia Book? It would have been helpful to be able to pick out the part which

> he actually contributed. However, a letter appealing for information in this connection, and sent to the only address given in the book, namely "Urantia Foundation, Chicago, Illinois," was returned marked "Address inadequate". So one is left wondering.

It was only when the Shermans' Urantia diaries were made available to the public in 2000, and when the Sherman-Loose correspondence was released soon afterwards, that the fuller story of Wilkins' Urantia connection could be told.

WILKINS' URANTIA NOTEBOOK

After Sir Hubert's death in 1958, the Notebook remained among his vast collection of papers, films and artifacts, in the custody of his widow, Suzanne. Upon her death in 1975, the Wilkinses' longtime secretary, Winston Ross, and his wife, Marley, inherited the collection. They were also bequeathed Walhalla, the Wilkinses' 107-acre farm in Pennsylvania, where the collection was stored. In 1985 Winston adopted Marley's son, Mike, and gave him "right of survivorship" after their deaths, to protect what would be left of the estate at that time.

Sometime in the 1980s the Rosses decided to create a Sir Hubert Wilkins museum. In order to finance it, they made a deal with two collectors of polar memorabilia to share the profits from the sale of Sir Hubert's vast collection of stamps, covers and letters. When Winston learned that most of those items went into the personal collections of the two men, he sued for an accounting. The collectors countersued. To prevent the possibility that the Rosses might lose the farm and everything stored on it, Mike Ross paid off the lawyers and a settlement was agreed upon wherein the Byrd Polar Research Center, at Ohio State University, in Columbus, Ohio, would receive the memorabilia. Winston Ross, however, held back items he could not bear to part with; other items, according to Mike, were lost. Winston Ross died in 1995 and Marley in 1997. Shortly before Winston's death, Mike rescued a number of boxes and stored them at his home in Michigan, where they remained until recently.

In the early 1960s both Mike and his mother became students of the Urantia Book, having discovered the copy Wilkins had given to Winston, who had barely opened it. All agreed that it would have been Sir Hubert's wish to donate Walhalla to the Urantia Foundation and the Urantia Brotherhood. In 1988 Mike negotiated with their representatives, but the organizations declined the offer.

In 2002 we (Saskia and Matthew) published the first volume of *The Sherman Diaries*, which featured Wilkins. Mike had heard about the book through the Internet and contacted me (Saskia) to find out what I knew about Wilkins' involvement with Urantia, as he himself had only sketchy details. Mike also mentioned that he had some Wilkins papers and other items in storage. I asked him about the Notebook, and although he remembered seeing it at the farm in the 1980s, he did not recall it being included in his boxes. We reluctantly concluded that most likely it was lost forever.

In 2014, as I was piecing together *Explorers of the Mind*, the correspondence between Sir Hubert and Harold Sherman, I tracked Mike down again to inform him of my project and to ask if he would allow me to rummage through his boxes for missing letters. He agreed. In preparation for my visit in October 2014, Mike opened the boxes for the first time in years, and spotted the lost Notebook almost immediately! Needless to say we were both thrilled.

Shortly before his passing on 30 November 2016, Mike donated his Wilkins collection, including the Notebook, to the Byrd Polar and Climate Research Center to join the bulk of Wilkins' papers and artifacts. Before that happened, he graciously allowed us to take the Notebook home to copy and transcribe all 356 pages, which we have since published in our book, *The Urantia Notebook of Sir Hubert Wilkins*.

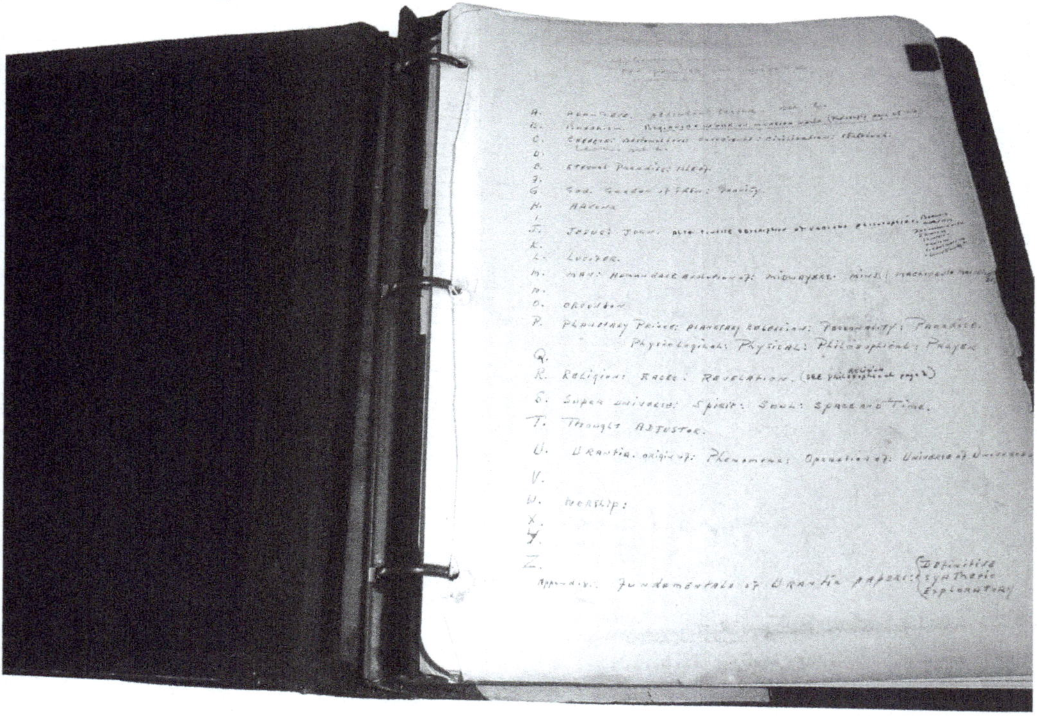

The black notebook that was missing for many years [MR]

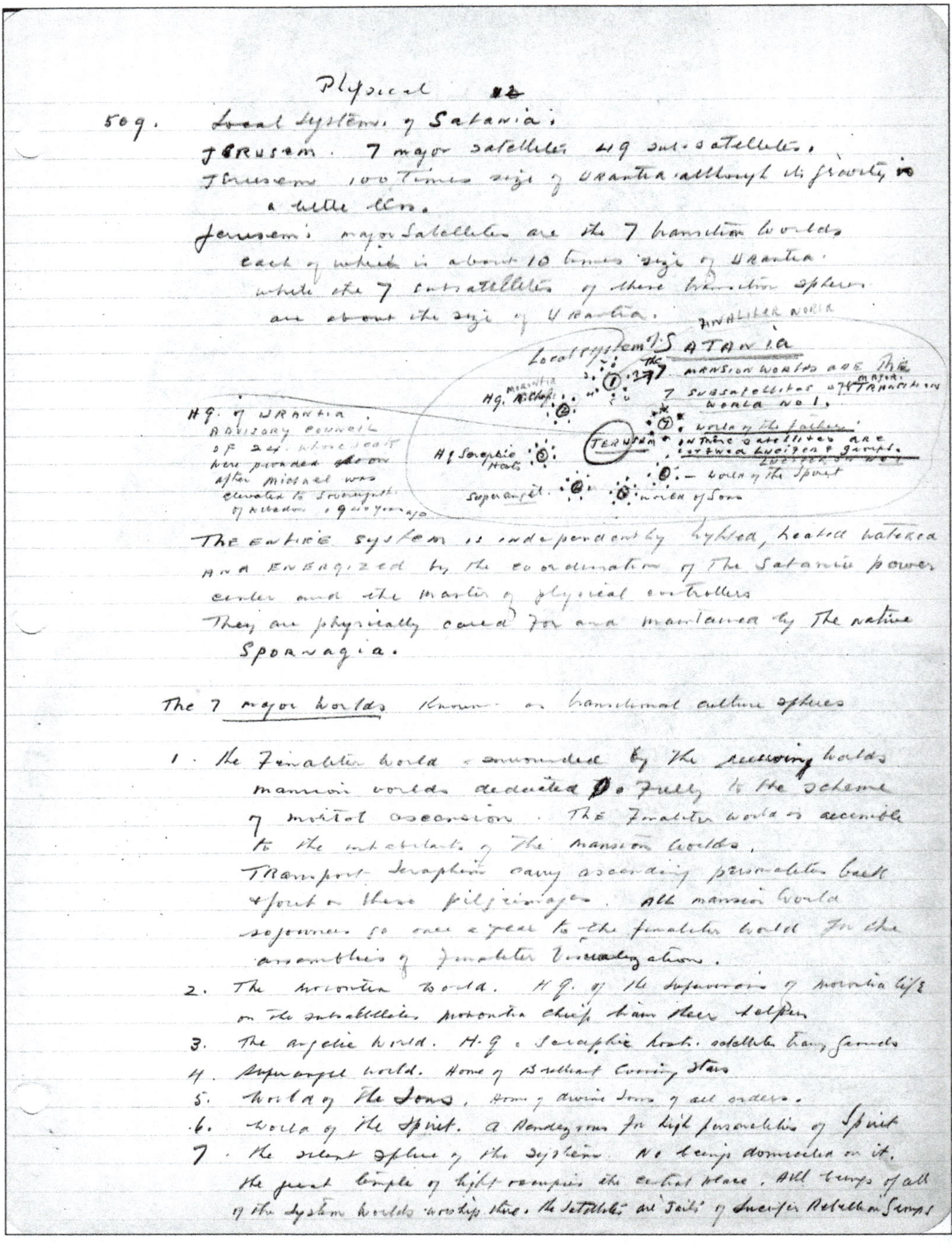

Physical #2

509. Local Systems of Satania:
Jerusem. 7 major satellites 49 sub-satellites.
Jerusem 100 times size of Urantia although its gravity is a little less.
Jerusem's major satellites are the 7 transition worlds each of which is about 10 times size of Urantia. while the 7 subsatellites of these transition spheres are about the size of Urantia.

The entire system is independently lighted, heated, watered and energized by the coordination of the Satania power center and the master physical controllers
They are physically cared for and maintained by the native Spornagia.

The 7 major worlds known as transitional culture spheres

1. The Finaliter World, surrounded by the receiving worlds mansion worlds dedicated fully to the scheme of mortal ascension. The Finaliter world is accessible to the inhabitants of the mansion worlds. Transport seraphim carry ascending personalities back & forth on these pilgrimages. All mansion world sojourners go once a year to the finaliter world for the assemblies of finaliter [illegible].
2. The Morontia World. H.Q. of the supervisors of morontia life on the subsatellites morontia chiefs [illegible] helpers
3. The angelic world. H.Q. Seraphic hosts. satellites [illegible]
4. Superangel world. Home of Brilliant Evening Stars
5. World of the Sons. Home of divine Sons of all orders.
6. World of the Spirit. A Rendezvous for high personalities of Spirit
7. The silent sphere of the system. No beings domiciled on it. The great temple of light occupies the central place. All beings of all of the system worlds worship there. The satellites are "Jails" of Lucifer Rebellion Groups

A page from Wilkins' Urantia notebook [MR]

Physical 12

509. Local System of Satania.
Jerusem 7 major satellites 49 sub-satellites
Jerusem 100 times size of Urantia although its gravity is a little less.
Jerusem's major satellites are the 7 transition worlds each of which is about 10 times size of Urantia - while the 7 subsatellites of these transition spheres are about the size of Urantia.

HQ of Urantia Advisory Council of 24 whose seats were provided soon after Michael was elevated to sovereignty of Nebadon 1900 years ago

<ILLUSTRATION>

The entire system is independently lighted, heated watered and energized by the coordination of the Satania power center and the master of physical controllers [*sic*].
They are physically cared for and maintained by the native spornagia.

The 7 major worlds known as transitional culture spheres

1. The Finaliter World - surrounded by the receiving worlds mansion worlds dedicated so fully to the scheme of mortal ascension. The finaliter world is accessible to the inhabitants of the mansion worlds. Transport seraphim carry ascending personalities back & forth on these pilgrimages. All mansion world sojourners go once a year to the finaliter world for the assemblies of finaliter visualization.
2. The Morontia World. H.Q. of the supervisors of morontia life on the subsatellites morontia chiefs train their helpers
3. The angelic world. H.Q. of the seraphic hosts - satellites training grounds.
4. Superangel world. Home of Brilliant Evening Stars
5. World of the Sons. Home of divine Sons of all orders.
6. World of the Spirit. A rendezvous for high personalities of Spirit
7. The Silent Sphere of the System. No beings domiciled on it, the great temple of light occupies the central place. All beings of all of the System Worlds worship there. The satellites are 'jails' of Lucifer rebellion groups

NOTES

1 Sir George Hubert Wilkins, *Under the North Pole: The Wilkins–Ellsworth Submarine Expedition* (New York, Brewer, Warren and Putnam, 1931), p. 73.
2 ibid., p. 82.
3 ibid., pp. 82-83.
4 This is the account Wilkins gave in *Under the North Pole.* In another version, dramatized for the radio program *True Adventure Thrills,* Wilkins wrote that he stowed away and was put off the boat in Algiers, where he was kidnapped by gun runners and rescued by an Arab girl.
5 Bernard Grant, together with Philip Gibbs who had been with the Bulgars, wrote a book about the war as viewed from both the Turkish and Bulgarian sides. It was published in 1913 as *Adventures of War with Cross and Crescent.* Wilkins contributed heavily and is mentioned in many of the stories. https://archive.org/stream/balkanwaradventu00gibbuoft#page/n7/mode/2up.
6 Lowell Thomas, *Sir Hubert Wilkins: His World Of Adventure* (New York, McGraw-Hill, 1961), p. 60.
7 John Grierson, *Sir Hubert Wilkins, Enigma of Exploration* (London, Robert Hale, 1960), p. 55.
8 Simon Nasht, *The Last Explorer: Hubert Wilkins—Australia's Unknown Hero* (Sydney, Hodder Australia, 2005), p. 55.
9 John Grierson, *Sir Hubert Wilkins, Enigma of Exploration* (London, Robert Hale, 1960), p. 60.
10 ibid., p. 64.
11 ibid., p. 65.
12 Simon Nasht, *The Last Explorer: Hubert Wilkins—Australia's Unknown Hero* (Sydney, Hodder Australia), p. 86
13 Malcolm Andrews, *Hubert Who?* (HarperCollins Publishers. Kindle Edition, 2011), Location 1374.
14 Lowell Thomas, *Sir Hubert Wilkins: His World Of Adventure* (New York, McGraw-Hill, 1961), p. 131.
15 ibid., p. 141.
16 Simon Nasht, *The Last Explorer: Hubert Wilkins—Australia's Unknown Hero* (Sydney, Hodder Australia,), p. 108.
17 British Film Institute: http://explore.bfi.org.uk/4ce2b6a25d482.
18 Sir Hubert Wilkins, "Mark Twain was Dead Wrong," unpublished manuscript, Byrd Polar and Climate Research Center, Ohio State University_Wilkins 22_22.
19 Malcolm Andrews, *Hubert Who?* (HarperCollins Publishers. Kindle Edition, 2011), Loc. 1741.
20 John Grierson, *Sir Hubert Wilkins, Enigma of Exploration* (London, Robert Hale, 1960), p. 77.

21 Simon Nasht, *The Last Explorer: Hubert Wilkins—Australia's Unknown Hero* (Sydney, Hodder Australia), p. 142.

22 Though the U.S. Navy and the National Geographic Society confirmed Byrd's achievement, experts who have studied his navigational records believe he fell short of the North Pole by as much as eighty miles.

23 Renamed the *Southern Cross,* Charles Kingsford Smith made history with the plane as the first to fly across the Pacific from the United States to Australia.

24 Lowell Thomas, *Sir Hubert Wilkins: His World Of Adventure* (New York, McGraw-Hill, 1961), p. 212.

25 ibid., p. 200.

26 Simon Nasht, *The Last Explorer: Hubert Wilkins—Australia's Unknown Hero* (Sydney, Hodder Australia), p. 17.

27 ibid., p. 168.

28 Lowell Thomas, *Sir Hubert Wilkins: His World Of Adventure* (New York, McGraw-Hill, 1961), p. 225.

29 In 1940 members of the United States Antarctic Service determined that Hearst Land was actually an island. They renamed it Wilkins Island.

30 Simon Nasht, *The Last Explorer: Hubert Wilkins—Australia's Unknown Hero* (Sydney, Hodder Australia), p. 187.

31 Lowell Thomas, *Sir Hubert Wilkins: His World Of Adventure* (New York, McGraw-Hill, 1961), p. 246.

32 Eielson had taken a job setting up Alaskan Airways and hired Joe Crosson as his chief pilot. Eielson died on November 9, 1929, in Siberia, when his plane crashed while attempting to evacuate a cargo vessel trapped in the ice. Wilkins was devastated by the news. He wrote: "Whatever else we may have accomplished through our efforts, we will only learn as time goes on. Eielson and I have learned, at all events, the sincerity of friendship."

33 "The Loneliest Wife in the World," Suzanne Wilkins, manuscript undated, circa 1937, Byrd Polar and Climate Research Center, Ohio State University_Wilkins, 21_12.

34 http://trove.nla.gov.au/ndp/del/article/29799911.

35 Simon Nasht, *The Last Explorer: Hubert Wilkins—Australia's Unknown Hero* (Sydney, Hodder Australia), p. 249.

36 Lowell Thomas, *Sir Hubert Wilkins: His World Of Adventure* (New York, McGraw-Hill, 1961), p. 270.

37 Today the *Nautilus* sits underwater near Bergen and is a popular dive destination.

38 One year later the *Hindenburg* tragically exploded over Lakehurst, bringing an end to the airship era.

39 Email Mike Ross to Saskia Praamsma, February 26, 2015.

40 John Grierson, *Sir Hubert Wilkins, Enigma of Exploration* (London, Robert Hale, 1960), p. 198.

41 ibid., p. 199.

42 ibid., p. 199.

43 Correspondence, Sir Hubert Wilkins to Harold Sherman, 6/20/42.

44 ibid., 5/13/46.

45 ibid., 10/14/46.

46 ibid., 12/31/46.

47 ibid., 6/18/52.

48 ibid., 9/8/52.

49 ibid., 6/1/58.

50 ibid., 3/23/56.

51 Correspondence, Suzanne Wilkins to Harold Sherman, 3/9/60.

52 "Sir Hubert Wilkins," by Elder Stanford G. Smith. The Church of Jesus Christ of Latter-day Saints, South Africa, *Cumorah's Southern Messenger*, September 26, 1933. https://archive.org/stream/cumorahsouthern0710eng/cumorahsouthern0710eng_djvu.txt.

53 Correspondence, Sir Hubert Wilkins to Suzanne Wilkins, collection of Mike Ross now housed at Byrd Polar and Climate Research Center, Ohio State University.

BIBLIOGRAPHY

Sir George Hubert Wilkins, *Under the North Pole: The Wilkins–Ellsworth Submarine Expedition* (New York, Brewer, Warren and Putnam, 1931)

Sir Hubert Wilkins and Harold Sherman, *Thoughts Through Space* (Hollywood, House-Warven, second edition, 1951)

Anonymous, *The Urantia Book* (Chicago, Urantia Foundation, 1955)

John Grierson, *Sir Hubert Wilkins, Enigma of Exploration* (London, Robert Hale, 1960)

Lowell Thomas, *Sir Hubert Wilkins: His World Of Adventure* (New York, McGraw-Hill, 1961)

Saskia Praamsma and Matthew Block, *The Urantia Diaries of Harold and Martha Sherman, Vols. I-VI* (Pahrump, Nevada, Square Circles Publishing, 2016-2021)

Simon Nasht, *The Last Explorer: Hubert Wilkins—Australia's Unknown Hero* (Sydney, Hodder Australia, 2005)

Jeff Maynard, *Wings of Ice* (North Sydney, Australia, Random House, 2010)

Malcolm Andrews, *Hubert Who?* (Sydney, HarperCollins Publishers, 2011)

ACKNOWLEDGMENTS

Thanks go first and foremost to the late Mike Ross for not only finding Sir Hubert's long-lost Urantia notebook and suggesting we take it home to copy and transcribe it, but also for allowing me to delve into the many boxes of Wilkins' papers in his possession before he donated them to the Byrd Polar and Climate Research Center at Ohio State University in Columbus.

To Laura J. Kissel, Polar curator at BPCRC, who was most welcoming and helpful during my visit.

To Jimmy Bryant and the staff at the Archives and Special Collections at the University of Central Arkansas, where the Harold Sherman papers are housed, for opening all the doors for me during my many visits.

To Mary Kobiella and Marcia Lynch—the daughters of Harold and Martha Sherman—who went extra miles in giving us access to documents and photographs still in their possession, before they were transferred to the UCA archive.

To our fellow Wilkins enthusiasts in Australia, Jeff Maynard and Stephen Carthew, who were most generous in sharing information about Sir Hubert, our common motive being to bring our forgotten hero into the limelight.

Saskia Praamsma
Matthew Block

INDEX

N

O

P

Q

R

S

T

Saskia Praamsma and Matthew Block are longtime Urantia Book readers who produce and distribute metaphysical, biographical and Urantia-related works via their company, Square Circles Publishing, and their websites:

squarecirclespublishing.com
squarecircles.com
urantiabooksources.com.

THE URANTIA DIARIES
of Harold and Martha Sherman

VOLUME 1: 1898-1942

ISBN: 978-0-9967165-9-8
eISBN: 978-0-9982556-0-6

VOLUME 2: 1942

ISBN: 978-1-7321796-1-5
eISBN: 978-1-7321796-2-2

VOLUME 3: 1943

ISBN: 978-1-7321796-3-9
eISBN: 978-1-7321796-4-6

VOLUME 4: 1944-1945

ISBN: 978-1-7321796-5-3
eISBN: 978-1-7321796-6-0

VOLUME 5: 1946-1955

ISBN: 978-1-7321796-7-7
eISBN: 978-1-7321796-8-4

VOLUME 6: 1955-1998

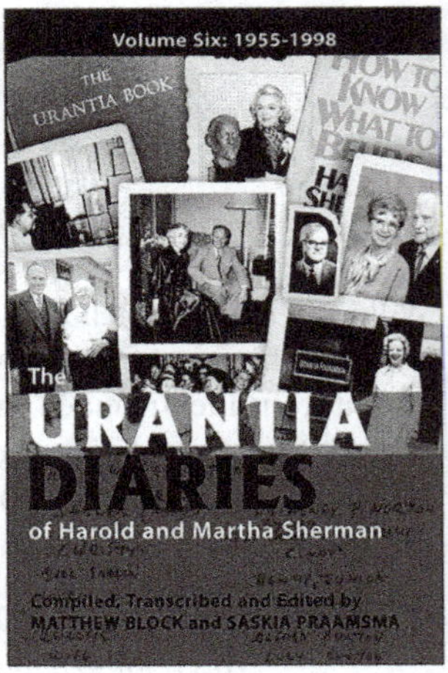

ISBN: 978-1-7344621-3-5
eISBN: 978-1-7344621-2-8

Available in paperback and Kindle at Amazon.com

Made in the USA
Monee, IL
04 November 2026

67380091R00079